Not by Fire but by Ice

Robert W. Felix

Not by Fire but by Ice

Discover What Killed the Dinosaurs... and Why it Could Soon Kill Us

Robert W. Felix

SUGARHOUSE
PUBLISHING
Bellevue, Washington

Cover graphic design by Virginia Hand
Cover illustration by Carol Stecher Jones
Author photo by Yuen Lui

Library of Congress Catalog Card Number: 96-70941

Publisher's Cataloging-in-Publication Data

Felix, Robert W.
    Not by Fire but by Ice:
    Discover What Killed the Dinosaurs
    . . . and Why it Could Soon Kill Us

    Includes bibliographical references and index.

        ISBN 0-9648746-9-5

        1. Ice ages—Cycle. 2. Geomagnetic reversals—Cycle.
        3. Extinctions—Cycle. 4. Paleontology.

Printed in Canada
c 10 9 8 7 6 5 4 3 2 1

Sugarhouse Publishing, P.O. Box 435, Bellevue, WA 98009

A long time ago,
the universe was made of ice.
Then one day
the ice began to melt,
and a mist rose into the sky.
Out of the mist
came a giant made of frost,
and the earth and the heavens
were made from his body.
That is how the world began,
and that is how the world will end,
not by fire
but by ice.
The seas will freeze,
and winters will never end.
—ANCIENT SCANDINAVIAN LEGEND

CONTENTS

Preface

LIST OF ILLUSTRATIONS

ACKNOWLEDGMENTS

So many terrific people gave me help and encouragement as I wrote this book that it's hard to know who to thank first. But I'll begin with Peggy King Anderson. Thanks, Peggy, for being so supportive . . . and for getting me hooked on writing in the first place.

And thanks to those brave souls who waded through so many "final" drafts. Heartfelt thanks to Lou Guzzo, S. Warren Carey, Mary Young, Douglas Gantenbein, Marion Terrell, Brian Gere, Dale Hamilton, Judy Sorrell, Lisle Rose, and to my daughter Michele, for their thoughtful comments and feedback. And special thanks to Leslee Tessmann and Phyllis Hatfield for their insightful and invaluable editing.

Most of all, I want to thank my long-suffering sister and brother-in-law, June and George Mona, who began plowing through the manuscript five long years ago when it consisted of 420 excruciatingly slow and boring pages. Then they read the 370-page version. And the 355-page version. And the 300 pager. And the 270 pager, and on, and on, and on. Thank you, June and George, for a real labor of love.

PREFACE

Let me tell you what this book is not. It is not another horror story about how we greedy humans are destroying our world. It's about mass extinctions and why the next mass extinction—including our own—could strike any day.

Pick up almost any newspaper and you'll read that our ecosystem is falling apart. You'll read that methane levels in our skies, along with carbon dioxide, hydrocarbons, sulfur, and nitrogen oxides, are rising, that ozone levels are dropping, and that temperatures in our seas are soaring. Those changes are caused, the story will swear, by humans.

Nothing could be further from the truth.

Most of those changes are caused by a naturally recurring cycle of earthquakes, volcanoes, floods, warming seas, mass extinctions, and glaciation.

We're not destroying our world; our world is about to destroy us.

While we sit here stewing about global warming, the next ice age is about to begin. And when it begins, it will begin with a bang! *Every* ice age began fast. One day you'll wake up—or won't wake up, rather—buried beneath nine stories of snow. By the time you finish this book, you'll shudder every time you see a snowflake.

Ice ages begin or end abruptly every 11,500 years. First comes an enormous flood, a Noah's Deluge-type of flood, which ends the previous ice age. Then comes a period of warmth similar to today's. This period of warmth, called an interglacial, lasts about 11,500 years. Then the next ice age begins . . . catastrophically.

All previous ice ages, let me repeat, began abruptly.

This 11,500-year cycle of warmth followed by an ice age has returned like clockwork for millions of years. To hope it won't happen
again just because humans now inhabit this planet would be wishful
thinking.

And that brings us to the crux of the matter. Since every period
of warmth—every interglacial—lasted almost exactly 11,500 years,
and since today's period of warmth began almost exactly 11,500
years ago . . .

You don't need a Ph.D. to understand this book. You don't need
to be a scientist, or a geologist, or a paleontologist, or any other
kind of an "ist." All you need to be is concerned with your own survival. It's written in laymen's terms. It's also written to be read fast.
If you don't understand something, just rush on through. It will become clear as you go.

Every part of this book, every word, leads to one inescapable
conclusion—that the next ice age could begin tomorrow. To reach
that conclusion, we begin with the dinosaur extinction of 65 million
years ago. Indeed, that's what got me hooked in the first place. I've
had an abiding interest in the dinosaur extinction and have devoured
every scrap of information on their demise that I could find.

What mysterious force of nature, I wondered, could kill so many
things all at once? The dinosaurs didn't slink off to their graves all
alone, you see, they took half the world with them. More than half.
Mammals, fish, trees, even plants, took a hit. Seventy-five percent
of all living species disappeared from the face of the earth in a
geological eye-blink. But why?

Along came the asteroid (or meteor) theory, proposing that a headlong collision between the earth and an asteroid killed the dinosaurs.
It was an exciting theory, easily visualized, and quickly became the
theory *du jour*. It still is.

However, many paleontologists, more than you might believe, disagree with the asteroid theory. So do I.

If the dinosaur extinction were the only mass killing to have ever
occurred on our planet, maybe I could buy the extinctions-are-
caused-by-an-asteroid theory. But it wasn't the only one. Our world
has endured at least 17 mass extinctions through the years.

But if an asteroid didn't kill the dinosaurs, what did?

One day—February 4, 1991, to be exact—all of the tiny tidbits of information that I had accumulated over the years came together in one epiphanic flash. I suddenly realized what really killed the dinosaurs . . . and that the same thing could soon kill most of us.

I felt compelled—driven, if you will—to follow that flash. Today, after five and a half years of full-time research and writing, I'm more convinced than ever.

Oddly, though, my original insight had nothing to do with ice. It had to do with geomagnetic reversals.

The earth's magnetic field has reversed itself several hundreds, perhaps thousands, of times over the course of history. During those times, compasses (if they'd been invented) would have pointed south instead of north, toward Antarctica instead of toward the Arctic. (Don't misunderstand me here, the world didn't tip upside down or anything like that. It's just that the magnetic north pole moved to the bottom of the globe.)

By some grand "coincidence," the dinosaur extinction occurred at one of those reversals. Most scientists think the reversal would have caused "no meaningful consequences."

I think they're woefully wrong.

Magnetic reversals are far more deadly than we ever dreamed. Magnetic reversals cause extinctions. Magnetic reversals trigger earthquakes and volcanoes. Magnetic reversals drive mountains into the sky. Magnetic reversals shoot electricity through the soil. And magnetic reversals cause rising and falling seas.

A magnetic reversal killed the dinosaurs, I will show, a magnetic reversal killed the mammoths, and a magnetic reversal will soon kill most of us.

Wait a minute. Didn't I say that we're going to die in an ice age?

Yes. But as unlikely as it may seem, magnetic reversals cause ice ages. In this book, you will learn how it's done. (Hint: It has to do with underwater volcanism.)

Ominously, the next magnetic reversal may have already begun.

All previous magnetic reversals were preceded by declining geo-magnetic field intensity. And today? Geomagnetic field intensity is falling. During the past 2,000 years, geomagnetic field intensity has plummeted more than 50%. Five percent of the decrease occurred in the last 100 years alone. This decrease, say experts, *may be a precursor to a new reversal attempt.*

And that's why I wrote this book—to warn you. We're sitting on the knife-edge of disaster . . . and we don't even know it. My hope is that everyone—professors, students, scientists, business leaders, government leaders, NASA, NOAA, the USGS, *everyone*—can pull together in search of survival solutions.

For Ashley Michele

*The age of reptiles ended because
it had gone on long enough and it
was all a mistake in the first place.*

—WILL CUPPY

1

· · · · · · ·

K-O'd at the K-T

· · · · · · ·

Contorted and twisted, their bodies lie scattered across the barren Wyoming countryside like dead flies on a window sill. But these are no flies, these are dinosaurs. Measuring up to 80 feet long (the length of a trailer-truck), the once mighty brontosaurs now lie embedded in the cold western limestone of Sheep Creek.

Arched backwards, with their necks and tails twisted and curved above their backs, the 50-ton monsters look as if they died in agony.

It must have been a muscle-pulling, head-twisting, bellowing kind of agony, to leave them so distorted and bowed. Dozens of similarly contorted bodies have been found in bone quarry after desolate bone quarry all over the west.

Nearly entire mummified dinosaur bodies, lying on their backs in awkward poses with all four feet stretched into the air, and with their chests expanded, have also been found at Lance Creek, Wyoming. They look, said paleontologist Robert Bakker, as if they died in one last gasp for breath.

Dinosaurs weren't the only animals that died. Their predators and scavengers died, too. They must have died at the same time, say paleontologists, because entire dinosaur carcasses have been found neither chewed nor ripped apart . . . and no self-respecting scavenger would turn up its nose at a free lunch.

What killed those multi-ton beasts, and their predators, and their scavengers? We've been asking that question since the first dinosaur fossil was discovered in 1822, but we still don't have the foggiest notion why.

Maybe the climate changed. Maybe it got too cold. Maybe it got too hot. Maybe there was too much rain. Maybe there wasn't enough. Maybe there was a drought. Desiccated [dried out] muscles and back ligaments, said Bakker, in his 1986 book *The Dinosaur Heresies*, could have pulled the dinosaur's heads and tails over their bodies precisely like that.

Maybe there are too many maybes. If it was a drought, why do we find deep dinosaur footprints—deep enough to hold 60 gallons of water—embedded in the same limestone? How do you make deep footprints in solid rock?

It must have been muddy the day they died, say geologists. The footprints were made before the limestone hardened. And since the footprints are deep, the mud must have been deep too.

Deep mud? During a *drought*?

How does a drought in Wyoming kill dinosaurs a thousand miles south in Texas? How does a drought in Wyoming kill dinosaurs 700 miles north in Alberta? How does a drought in Wyoming kill dinosaurs 2,000 miles east in the Connecticut River Valley? Or in

Africa? Or in Belgium? Dinosaur bones have been found all over the world, from Saudi Arabia to Antarctica. Did the so-called drought affect the entire planet?

And how does a drought in Wyoming kill fish thousands of miles away in the ocean? Oh, you didn't know about the fish? Dinosaurs didn't slink off to their graves all alone, you know; they took more than half the world with them. Mammals, fish, trees, even plants, took a hit. Ocean-dwelling organisms were almost annihilated.

It was a mass extinction, global and sudden. Seventy-five percent of all species on the planet went extinct, never again to appear in the geologic record. The sheer number of other deaths, say scientists, make the dinosaurs' disappearance look like an afterthought.

When 75% of all living species disappear from the face of the earth, we've got a disaster on our hands, a disaster greater than any nuclear holocaust we've ever tried to imagine. If we simultaneously exploded every nuclear weapon in existence in every country on earth, say scientists, we would not begin to match the devastation. Not even close. There must be an answer, and we'd better find it quick, before it happens again.

It is going to happen again, and soon!

It's not as if the dinosaurs were newcomers and didn't know what they were doing. They had smarts. They were survivors. They'd been hanging around Cretaceous swamps for millions of years.[1] (The first Cretaceous fossils were found buried in chalk; "chalk" in Latin is *creta*, hence the name Cretaceous.)

Dinosaurs burst onto the scene about 248 million years ago at the beginning of the Mesozoic Era. They arrived as they left—in a hurry —like a conqueror in the night. No one knows where they came from. Quickly seizing power, they soon dominated the world, reigning over their newfound kingdom for more than 180 million years. It looked as if they would live forever.

---

1. Dinosaurs didn't really live in swamps, by the way, as so many children's books try to depict them. It's just that when they died, gigantic floods swept their bodies into ancient rivers and lakes, making it look as if that's where they had lived.

Compared to the dinosaurs, our time on earth looks like a period at the end of a book. We're babies. If you made a chart showing how long the dinosaurs ruled the world compared to us, and made our side of the chart a mere one inch tall, the dinosaur's side would soar 150 stories into the sky.

King of the hill, robust and healthy, rulers of all they surveyed, there is no hint in the fossil record of their impending demise. But then, wham!—with no warning, no prior thinning of their ranks, no time to make peace with their gods, no time, even, for unhatched dinosaur babies to emerge from their shells, they were brutally murdered. The street fighters of their time, why didn't they battle for their lives? No one knows.

Their graves mark the end of one geological period and the start of another; the end of the Cretaceous Period and the start of the Tertiary. The 65-million-year-old dividing line between the two periods is called the K-T boundary. (The "K" comes from *kreide*, the German word for *creta*. And the "T" stands for Tertiary, which means third.) It marks the beginning of the third major era in geologic history.

Paleontologists have searched the rocks of past geologic periods in every way they know how. They've searched from horseback, they've searched from satellites, they've searched from planes, and they've crawled over scorched desert rocks on their hands and knees. They've used dynamite, they've used jackhammers, they've used feather dusters and dental drills. No matter. To this day, no one agrees as to what killed those awe-inspiring reptiles.

Stress killed the dinosaurs, one scientist proposed, after learning that dinosaur eggshells had grown thinner over time. Thinner eggshells prove that the dinosaurs had become fat, lazy, and stressed out.

Seems like an odd combination, doesn't it? But listen to his reasoning. The luxurious swamps of the end-Cretaceous and the warm climate, he theorized, gave the dinosaurs a fat, easy life. Like a perpetual Bacchanalian party, the happy dinosaurs simply rolled over, opened their mouths and waited, while food practically fell from the sky.

Fast-food to the max! Life was so easy, all they did was eat and have sex. That led to overpopulation and overcrowding. Which, just as it does with people, inevitably led to stress.

Stress kills birds. Or rather, it interrupts their egg production, and then they die off. (Stress creates an imbalance in their hormonal system, making their eggshells too thin for proper hatching.) Maybe the same kind of stress got to the overcrowded dinosaurs.

Another theory lays the blame on mammals. There *were* mammals back then. They were insignificant little things about the size of a rodent, probably nocturnal. "Fur balls," scientists call them. (Officially, they're called stem mammals.) Maybe our mammalian ancestors ate the eggs. What a life! Dinosaur omelettes every night for a midnight snack.

But plants don't lay eggs. Why did so many plants disappear? Angiosperms (flowering plants and hardwood trees) disappeared almost overnight. Magnolia trees, or something similar, disappeared in a single growing season, in four to six weeks, said Jack Wolfe of the U.S. Geological Survey (USGS). The only thing left were the ferns. It's called "the fern spike," and it proves the extinction came suddenly. It was an ecological St. Valentine's Day Massacre, said Stephen Jay Gould of Harvard. "The ecosystem totally collapsed."

Plant pollen and spores disappeared so fast at the boundary, said Robert Tschudy of the USGS, that "we can envision the forests dead in less than a year's time." And for some reason, he noted, more plants died in the north than in the tropics. (Leo Hickey at the Smithsonian also found more damage to the north.)

Maybe the plants themselves were the culprits, said Anthony Hallam, a well-known British paleontologist. Maybe the reptile's new diet upset their tummies. "One is led ineluctably to the conclusion," said Hallam, "that the poor dinosaurs died of constipation."

Or you may have seen the article in your local paper touting the thought that massive emissions of methane gas—caused by dinosaur flatulence—may have caused global warming, thereby doing whatever it is that global warming is supposed to do.

Maybe every glacier in the world melted at the same time, and the seas rose hundreds of feet above normal, flooding the continents.

No, say others, the oceans went down at the end-Cretaceous, not up. They fell an incredible fifth of a mile. No one knows why.

Meanwhile, many world-class geologists with rock-solid credentials think volcanoes killed the dinosaurs. It's a legitimate argument. Massive volcanic eruptions occurred all over the world right at the K-T boundary.

Maybe the dinosaurs had grown so huge, one wag suggested, that they could no longer mate. But he hadn't done his homework. Sure, some dinosaurs were monsters. But they weren't all the size of a five-story house. Some were as small as a dog, others were the size of a crow.

Even if they were big, why did so many other animals die? Sea dwellers big and small, from giant sea-serpents to rudistid clams, were almost decimated. And belemnites, though prolific, completely disappeared, taking many other apparently hardy mollusks with them. (Belemnites, relatives of today's squid, look like fossilized stone cigars; or "sea-gars," as the locals call them.)

Sea dwellers had not declined or lost vigor, said Surlyk and Johansen in *Science* (writing about Cretaceous lampshells). Healthy and numerous, there was no warning of their impending doom, neither in the form of decreasing population density nor in diversity. There was nothing about the bryozoans, they said, nor the foraminifera, nor the brachiopods, to indicate that their demise was near.

Ammonites, big players in the Cretaceous seas, were no luckier. When the extinction bell sounded, they were knocked from the ring like everyone else.[1] K-O'd at the K-T.

Suddenly, sediments turned almost totally barren. Even plankton were pushed to the brink of extinction. What could have killed so many things all at once? And do it so fast?

---

1. Ammonites, essentially octopi in a shell, lived in tightly coiled spiral shells and were closely related to today's pearly nautilus. Sometimes no bigger than a coin, other times larger than a tractor tire (more than 12 feet across), they flooded or pumped interior chambers to rise and fall through the water like miniature submarines. They were once thought to be the horns of Amon, the ancient Egyptian god.

Fast? Paleontologists used to think that mass extinctions, including the K-T, were long, drawn-out affairs occurring over hundreds of thousands of years; simply a bunch of background extinctions that kept adding up. Today, most scientists agree that the dinosaur extinction occurred blindingly fast.

They died in less than 1,000 years, say some paleontologists.

No, said Norman D. Newell of the American Museum of Natural History in New York, they died within a few hundred years.

No, said Jan Smit of the University of Amsterdam, they died in less than 50 years.

No, it was almost instantaneous, said Digby McLaren, a noted paleontologist at the University of Ottawa in Ontario.

Instantaneous. A dirty word. A word that paleontologists avoid like a root canal. But instantaneous seems to be what the fossil record is saying.

Maybe a comet did it, said the French scientist Pierre de Maupertuis. No, you didn't read about him in last week's paper. Far ahead of his time, de Maupertuis made his daring proposal back in 1750.

He was wrong, of course.

Present-day asteroid theories are wrong too, you will soon see.

Asteroids aren't choosey, right? So why were ocean dwellers almost annihilated, while freshwater animals survived? Crocodiles and alligators came through with flying colors. So did freshwater turtles. And freshwater invertebrates like mollusks in rivers and lakes, says paleontologist Kenneth Hsü, weren't affected at all. It's a mystery worthy of Sherlock Holmes. Freshwater somehow protected them.

Disparities occurred within the oceans, too. Animals living in deep water fared far better than those who lived in the shallows. No one knows why.

Those are just some of the mysteries.

On land, all animals weighing more than 55 pounds were killed, while many small animals survived. Many families of lizards and mammals passed through the disaster almost unfazed.

So what's the answer? When paleontologists try to understand the past, they look at the layers of soil, clay, and rock, and the fossils embedded in those layers, that were deposited during the period they're studying.

Finding a place that has soil of the proper age, in this case 65 million years old, is hard enough. But interpreting the clues hidden in that soil is an art, an art requiring the intuitive ability of a great detective. You need a sort of Inspector Clouseau, B.C.

Walter Alvarez was the detective, and Gubbio was the place. Gubbio, a small town in central Italy, has been famed since the 1700s for its ceramics made of clays from the nearby Apennine Mountains. Big-time secrets were trapped in those clays, waiting for the right detective to pry them loose.

They were heavy hitters, those detectives, a highly competent team from Berkeley. There was Walter Alvarez, a professor of geology at the University of California; his father Luis Alvarez, a physicist and Nobel prizewinner from Lawrence Berkeley Laboratory; and Frank Asaro and Helen Michel, two nuclear chemists from the same lab. Not a lightweight in the bunch.

But they were no dinosaur detectives.

Dinosaurs were the last thing on their minds on that fateful day in 1978. They didn't care about mosasaurs, or stegosaurs, or any other kind of saur. They were interested in periods of reversed polarity, times when compasses would have pointed south instead of north.[1]

Serendipity often plays a major role in new discoveries, and their interest in magnetic reversals would lead to one of the most significant discoveries in geologic history.

As Alvarez conducted experiments (neutron activation analysis) on the Gubbio clays, he discovered an anomaly, an irregularity. He found abnormally high concentrations, up to 30 times greater than normal, of iridium. (Iridium is a metallic element resembling platinum.) He also found unusually high concentrations of osmium,

---

1. Compasses would have pointed south during roughly half of history, said Allan Cox and Robert Hart in their 1986 book *Plate Tectonics, How it Works*. The earth's polarity has reversed hundreds, perhaps thousands, of times. The K-T extinction occurred at one of those reversals.

antimony, and arsenic. Waving an anomaly in a scientist's face is like giving him a fleeting glimpse of a striptease dancer. He wants to see more.

Alvarez peeked behind the curtain. Would he find iridium in other locations? A resounding yes! He found a half-inch-thick layer of iridium-laced clay all over the globe right at the K-T boundary. Like a bookmark between the pages of time, it was sandwiched between the dirty white limestone of the Cretaceous and the brownish limestone of the Tertiary.

Boundary clays from Spain, Italy, Texas, the North Pacific, the South Atlantic, it made no difference where they came from; all contained excess iridium. Boundary clays from New Zealand held 20 times normal concentrations. Clays from Denmark held even more.

Wherever boundary clays were discovered, they were invariably enriched with iridium. More than 100 scientists in 21 laboratories in 13 countries checking 95 different sites around the world found anomalously high levels of iridium. But how did such huge amounts of iridium get scattered all over the planet?

Though there seems to be plenty of iridium in the earth's core, it's rare on the crust. How did it get to the surface? And in a layer? It looked as if it had come from the sky.

Aha! It did come from the sky! Iridium is relatively common in celestial bodies such as asteroids and meteorites, Alvarez remembered. That dusting of iridium found around the earth must be from outer space. Extraterrestrial. But still, how did it get here?

He ruled out comets. They're just "big, dirty snowballs" made of frozen gases, frozen dust, and frozen pieces of rock. They could never contain enough iridium to blanket the entire planet.

Two years later, Alvarez dropped a bomb on the scientific establishment. The iridium points to a cataclysmic collision between the earth and a giant galactic intruder, he announced. A mountain-sized asteroid at least six miles in diameter had crashed into the earth head-on.

Crazy! A galactic solution to an earthly question. It was laughable. Alvarez had gone to Gubbio to study reversed polarity, and here he was, a geologist, suggesting that some sort of rogue asteroid had killed the dinosaurs. He had flipped. He should have stuck to his flipping polarity.

If only he had known . . .

But the scientific community didn't laugh. Instead, it rose to the challenge, embarking on a galactic-sized asteroid witch hunt. As reports trickled in from around the globe, Alvarez's theory looked like a winner. An impact would have caused havoc around the world, his fellow scientists agreed, setting off earthquakes, volcanoes, and God only knows what else.

If the asteroid had crashed into the ocean, said Alvarez, at least 1,250 billion tons of water would have shot into the air. Tremendous tsunami (tall waves) would have raced across the oceans at more than 400 miles an hour, building into walls of water half-a-mile high before they slammed into the unprepared continents.

Maybe higher! Alan Hildebrand and William Boynton at the University of Arizona think the waves could have reached an incredible three miles in height, rising far into the clouds. Spreading inland, they'd have destroyed everything in their paths. (This is not just an "if." Jody Bourgeois, a geological sleuth from the University of Washington, found evidence at the Brazos River near Waco, Texas, that a 65-million-year-old tsunami deposited a four-foot-deep layer of sand in just one day, and rolled Texas-sized boulders around as if they were pebbles.)[1]

Or, said Alvarez, an impact on land could have blasted a crater in the ground at least 20 times the size of the asteroid itself, a crater 125 miles across and 24 miles deep.

Debris thrown from the crater and spalled from the asteroid itself would have gone ballistic. Hurled violently into the atmosphere at up to 80,000 miles an hour, it would have charged around the planet in

---

1. Further proof of a tsunami comes from Cuba, where the Polish scientist A. Pszczolkowski found a 65-million-year-old graded bed about 450 meters thick (1,485 feet) apparently deposited by a massive wave. Another tsunami deposit of K-T age, topped by a layer of iridium, has been found in northeastern Mexico. (Smit *et al.*, 1992)

less than 20 minutes. Some rocks and pebbles, he envisioned, could have shot into orbit.

During the year following impact, as the ejected rocks and pebbles traced their trajectories around the planet, they'd have eventually succumbed to the laws of gravity. Reentering the earth's atmosphere, they'd have provided bright and deadly meteor showers of unprecedented proportions.

Tiny, burnt to a crisp, few of the projectiles would have reached the ground to leave any scars or permanent record of their short-lived existence. They'd have simply turned to dust and soot.

It seemed impossible.

But when University of Chicago graduate student Wendy Wolbach discovered a layer of carbon in boundary clays close to the iridium, asteroid believers were ecstatic.

There's so much carbon in the boundary clays, said Wolbach, that more than 90% of all vegetation on earth, and many millions of animals, must have burned as raging forest fires—ignited by the asteroid's impact—engulfed the entire planet. With every Cretaceous field and forest ablaze, how could any animal have survived?

Black and dirty, soot-filled skies would have turned day into night. Bewildered dinosaurs probably couldn't see their own tails. The darkness, combined with frigid temperatures (it gets cold in the dark), could have lasted for months, if not years. Chemical reactions could have formed nitric acid, sulfuric acid, and hydrochloric acid; acid rain, in other words, right in the air.

Acid rain, many scientists agree, could explain why so many kinds of plants and animals, both on the land and in the seas, were killed. They were swimming in toxic soup.

The impact could have triggered volcanic eruptions all over the globe, sending volcanic ash spiraling high into the atmosphere—and presto!—more acid rain.

And more darkness. A self-perpetuating disaster.

We're talking about the kind of damage you'd expect from a nuclear explosion. Not one of your piddling "everyday" nuclear explosions, either. Anything that would spread a half-inch layer of iridium-laced clay or more (it's a foot deep in places) over the entire globe had to have been caused by something catastrophic. We're talking about 10,000 times more damage, scientists calculate, than if we were to detonate every existing nuclear bomb in the world.

An impact in the sea, said Alvarez, would have thrown 60 times more material into the air than the actual mass of the asteroid itself. Up to sixty thousand billion tons of debris would have been ejected into the atmosphere, most of it from the ocean floor. All from an asteroid only six miles across.

The ocean would have exploded heavenward. As surrounding waters swooped in to fill the void and gushed into contact with the super-heated area of impact, tremendous plumes of steam and vapor would have shot skyward. A gigantic maelstrom of swirling air, sea-bed, water, steam, and pulverized fish would have risen into the skies like a tremendous nuclear mushroom cloud. With its stem more than 100 miles in diameter, the cloud itself would have been thousands of miles across, bigger than a continent.

Traveling at least 50,000 miles per hour (New York to Arizona in three minutes) the asteroid would have struck with so much force, Alvarez estimated, that it would have released a billion times more energy than the Hiroshima atomic bomb.

That wasn't a typo. Let me say it again. The impact would have been equivalent to the energy of one *billion* Hiroshima bombs!

It would have been like a long nuclear winter, with nothing left but a "Strangelove" ocean. By the time the dust settled, the world would have been a desolate barren wasteland. The immensity of the impact scenario fired the imaginations of excited scientists around the world, and they rushed forward with corroborating evidence.

Impact believers were on a roll, and by 1988 most scientists were convinced. "HUGE IMPACT IS FAVORED K-T BOUNDARY KILLER," heralded a story in a November 1988 issue of *Science*. Scientists give a "clear-cut victory" to an asteroid or cometary impact, the article burbled. It looked as if the issue was settled.

But they were wrong.

Believing that the dinosaurs went extinct as a result of asteroidal impact is a major miscalculation. Their deaths can be linked to climatic cooling.

—GERTA KELLER

2

· · · · · · · ·

BOLIDE BULLIES

· · · · · · · ·

If a meteor killed the dinosaurs, where's the hole? We're talking about a hole in the ground that's supposed to be 125 miles across—bigger than Connecticut, Rhode Island, and Delaware combined.

We're talking about a hole in the ground that's supposed to be 24 miles deep—eight times deeper than Mount Rainier is tall.

Where do you hide a hole that big? The asteroid must have crashed into the ocean, say impact believers, and now is buried in a watery grave. The water slowed it down so it made a smaller hole. Then the hole filled with sediments, making it harder to find. Or, they point out, it could have been swallowed into the earth by subduction. (Subduction is a process where the seafloor dives in slow motion beneath the continents like a train plunging into a tunnel. More later.)

Actually, two craters of about the right age do exist. The smaller of the two, the Manson Structure, lies beneath a blanket of glacial drift in Manson, Iowa. Though not exactly a pimple (it's 21 miles across), whatever made it would have been too small, say paleontologists, to have killed any dinosaurs unless they'd been standing directly in the line of fire.

The other, a 110-mile-wide crater named Chicxulub (chicks-oo-loob), lies south of the border on Mexico's Yucatán Peninsula. Chicxulub (the tail of the devil) was first discovered by workers at Pemex, the Mexican national oil company, when they drilled through a layer of andesite, a type of dark volcanic rock. (Swisher *et al.*)

Buried under more than half a mile of sediments, Chicxulub lies in a shallow part of the oil-rich Gulf of Mexico about 500 miles southeast of Brownsville, Texas. Chicxulub is proof, said Arizona's Hildebrand and Boynton, that an enormous meteorite struck the Caribbean at the end-Cretaceous.

But is Chicxulub big enough? No. Put Manson and Chicxulub together, say disappointed impact believers, and they're still too small. Unless we find more craters, they moan, we'll have to abandon the bolide theory. (The word bolide is used to describe comets, meteors, asteroids, or any other kind of astronomical object that might come flaming through outer space.)

Meandering? Misplaced? Eaten? Covered up? Where are those clever craters hiding? Maybe there aren't any. Many scientists, skeptical of the entire asteroid theory, keep looking elsewhere for the real cause of the extinction; for the real cause of all extinctions.

What about reversed polarity? After all, the earth's magnetic field did reverse at the time. That's why Alvarez went to Gubbio in the first place. Did the reversal cause the extinction? No, most scientists insist, a reversal would have caused "no meaningful consequences." The earth's magnetic field has reversed many times in the past, they point out, with no ill effects (that they know of).

But they have to say something, so they usually toss in an aside about some guy named Uffen who came up with a reversal theory back in the 1960s. Our magnetic field shields us from cosmic rays, said Uffen. When the field reversed, we temporarily lost our shielding, and mutation-causing cosmic rays bombarded the earth.

Cosmic rays? Far out. No one paid much attention to Uffen.

What a deadly mistake!

What else could have done it? The extinction was a purely volcanic event, said Charles Officer of Dartmouth College. Explosive volcanism occurred all around the globe right at the K-T boundary.

Others agree. "The very end of the Cretaceous was marked by an episode of paroxysmal [volcanic] activity," said Anthony Hallam of the University of Birmingham, U.K. (*Science*, 1987) But still, where did the iridium come from?

No problem. It came from volcanoes. Traces of iridium have been found in eruptions of acidic volcanoes on Russia's Kamchatka Peninsula, said John Lyons and Charles Officer.

The iridium need not have come from outer space, agreed volcanologist Vincent Courtillot. Emissions from Hawaii's Kilauea volcano contain iridium enhancements about 10,000 times more than normal. Iridium has also been found in ash from the Piton de la Fournaise, a volcano on the island of Réunion. Iridium-rich volcanic dust has even been found embedded in the Antarctic ice sheet, said Courtillot. If a handful of comparatively small volcanoes can spread iridium to the bottom of the globe, imagine what a large volcano like the Deccan Traps in India could have done.

More than two million cubic kilometers of molten death poured out of the Deccan Traps at the K-T boundary, smothering 220,000 square miles of western India (about a sixth of the country) under successive layers of lava up to a mile and a half deep. When the layers eroded, they left staggered cliffs in their wake. (That's where the name came from. *Trap* means *staircase* in Dutch, and *deccan* means *southern* in Sanskrit; the southern staircase.) Volcanoes must have been the culprit, say volcano believers. Dinosaur fragments are found immediately beneath the lava—but none are found above it.

Maybe the asteroid caused the Deccan Traps, impact proponents retorted, then was buried beneath the lava. But how would they explain the rift volcanism that occurred, at the same time, in the western United States, eastern Greenland, Great Britain, and Hawaii?

And how would they explain the Siberian Traps, which burped out millions of cubic kilometers of basalt some 248 million years ago at the end-Permian?

The granddaddy of all extinctions, the Great Permian Extinction was the largest mass extinction in history, far bigger than the K-T. Wiping out trees, plants, fish, plankton, animals, insects, even algae, it destroyed at least 96% of every kind of life-form on earth. Is there a meteor crater buried beneath the Siberian Traps?

It's doubtful. "There is no evidence of an asteroid or comet impact having anything to do with the P-T [Permian-Triassic] extinction," said Richard A. Kerr in *Science*. (26 Nov 1993)

Even if it did, what about the other volcano/extinction links? "Most major extinction events," said Courtillot, "correlate in time with large flood basalt eruptions." Some eruptions were so huge, he said, that their very size could have triggered the kind of destruction that Alvarez envisioned for the impact theory.

Besides, Chicxulub may not be a meteor crater at all. It may be volcanic. After all, it *was* discovered under a layer of volcanic rock. And the ejecta layer that Hildebrand and Boynton used in their studies (the layer of soil and rock supposedly ejected from the ground by the meteor strike) was originally accused of being volcanic by no less a figure than Florentin Maurasse, the geologist from Florida International University who discovered the ejecta layer in the first place.

The Haiti ejecta layer is probably volcanic, agree Lyons and Officer. "The vast bulk of the Beloc [Haiti] black glasses have compositions identical to those of volcanic rocks."

The evidence for volcanoes is persuasive. Then again, the evidence changes daily. In a 1991 lecture at the University of Washington, Dutch geologist Jan Smit announced the surprise discovery of yet a third crater candidate on the north coast of Siberia. Named Popigai, and measuring an impressive 80 miles across, the new crater lends enormous credence to the impact theory.[1] It's the "smoking gun" the impact people wanted, giving them enough craters to explain all the damage. "The extinction mystery is solved," said Smit, "the crater trio clinches the argument." Florentin Maurasse even switched sides. (Maurasse and Sen, 1991)

Impact believers were riding high. But then they got greedy. *All* extinctions were caused by meteors, they chanted. Standing on a mountain of speculation, they had jumped over the edge.

Look at the sheer number of mass extinctions that our planet has endured. (See the Geologic Time Scale on pages 38 and 39.)

We've had four mass extinctions (counting the mammoths) just *since* the dinosaurs died. One about 37 million years ago (mya) at the end-Eocene; one about 12 mya at the mid-Miocene; yet another about two million years ago at the end-Pleistocene. And the mammoths? They died a mere 11,500 years ago.

We're beginning to realize that earth is a violent and dangerous place to live. We're beginning to realize that mass extinctions have been the rule, rather than the exception, for the 3.5 billion years that life has existed on this planet. Almost identical, each extinction was abrupt, each was extensive, and each was caused by some temporary, unexplainable event. Are they saying that every mass extinction was caused by a meteor?

Well . . . sort of.

But wait until you hear what they came up with next. Extinctions don't strike at random, they said. Extinctions strike according to a recurring predictable cycle. How hare-brained can you get?

Uh-oh. This time they were right. There *is* a cycle. In the early 1980s, paleontologists J. John Sepkoski, Jr. and David M. Raup conducted extensive studies of the deaths of sea dwelling organisms. They found more extinctions than anyone had dreamed possible.

---

1. Actually, Popigai isn't "new." Paleontologists previously thought it was 40 million years old, but different dating techniques now place it at 65.

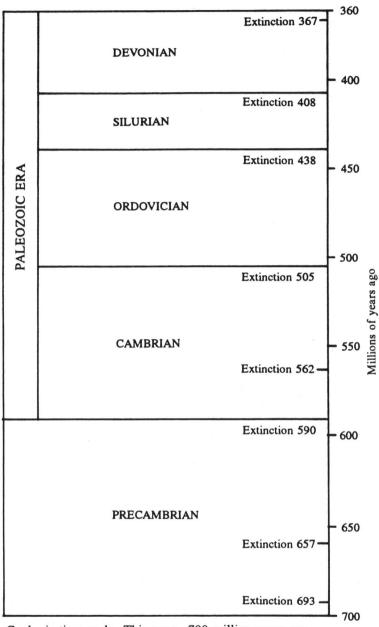

Geologic time scale. This page—700 million years ago
(mya) to 360 mya. Facing page—360 mya to present.

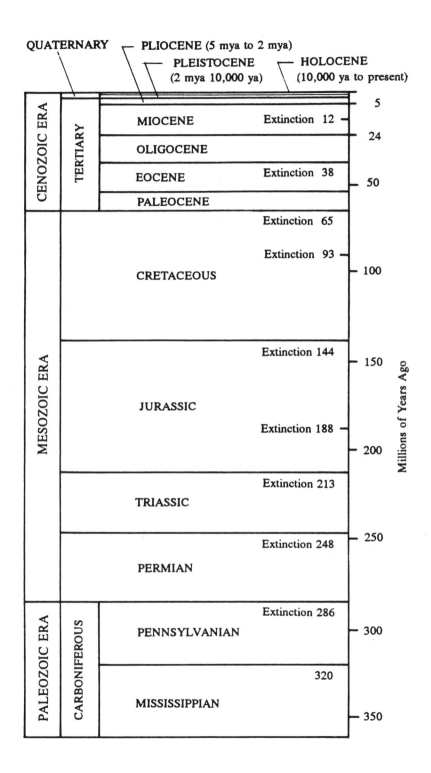

Taking all of their new-found extinctions, eliminating first one and then another (in case they were mistaken and it wasn't really an extinction), they crunched more than 1,000 possible configurations through their computers. A definite pattern popped out. Extinctions have occurred just like clockwork, they announced, every 26.2 million years. It was a major paleontological breakthrough.

What could cause such a cycle? They knew of no earthly process with a period of 26.2 million years. It must be something hidden in the solar system or galaxy.

Something extraterrestrial.

A cycle? Extraterrestrial? The father of the impact theory was intrigued. Is there a cycle to the creation of craters, too? Analyzing 16 craters at least six miles across, Alvarez and fellow scientist Richard Muller conducted their own experiments. There *is* a pattern, they agreed. But it's not a 26.2-million-year cycle. The craters were formed about 28.4 million years apart.

That's okay, said Raup and Sepkoski. When you're talking about something that old, 1.2 million years is close enough. Impact believers were ecstatic. There *is* an asteroid cycle, and it matches the extinction cycle!

Raup and Sepkoski didn't have a monopoly on cycle theories. According to Alfred Fischer and Michael Arthur in their 1977 paper *Secular Variations in the Pelagic Realm*, extinctions are evenly spaced at 32 million years apart. Hah! "I have seen better cycles in the Dow Jones averages," scoffed Raup.

Astronomers Michael Rampino and Richard Stothers of NASA also found a 29- to 31-million-year extinction cycle, along with a corresponding 31-million-year cycle to the creation of large craters. It must be connected, they decided, to our solar system's periodic trips through the galactic plane. We oscillate through the plane every 33 million years or so.

Others agreed. We bob up or down through the plane every 26 million to 37 million years, said John Bahcall of the Institute for Advanced Study in Princeton, N.J. "We last passed through the plane anywhere from now," he said, "to 5 million years ago, but probably about 3 million years ago." At present, we're lying 8° to 12° north, quite close to the plane.

What kind of diabolical killing machine is up there anyway? What could make the heavens open up and spit out asteroids every 26.2 million years, sending them crashing through space like clockwork?

Scientists didn't have a clue.

They needed an answer, and they needed it fast. So they invented one. But first, they had to invent a star. Dubbing it the "Nemesis Star," their theory goes something like this:

> The spiral arms of our galaxy contain planetesimals. Some of these planetesimals cluster together to form an area of interstellar clouds made of dust, gas, and comets called the Oort Cloud. As our solar system passes through the Oort Cloud—if it passes through it—perhaps it collides with a few of the comets.
>
> Or perhaps our sun has a shy companion star that no one has ever seen. (They had the audacity to invent another sun!) This companion star might have an unusual orbit that could take it through the galactic plane. If it should pass through the plane every 26.2 million years, or 28.4, or 32, or whatever, it might encounter that Oort Cloud.
>
> This could disturb the orbits of a few comets, thereby shaking enough of them loose so that one, or two, or maybe even three, could speed millions of miles through space and crash unerringly into the earth, at the same time, every 26.2 million years.

They think asteroids are a renewable resource, I guess, like fish, or trees, or fields of grain.

This hypothesis is taken quite seriously. I think it contains more "ifs," "coulds," "perhapses," and "maybes" . . . than comets.

For argument's sake though, let's pretend there really is an evil red companion star lurking up there, dodging behind other stars every time we try to find it, ready to attack the moment we drop our guard. If asteroid assassins strike the earth every 26.2 million years, where are all of the craters?

Where are the twelve craters (three per extinction) that should have been created since the K-T? And where are the hundreds of craters from before the K-T? If the earth has existed for 4.6 billion years, and if the Oort Cloud spits out three meteors every 26.2 million years, shouldn't we have been bombarded 525 times during the course of history?

We're able to find 400-million-year-old fossils smaller than a grain of sand—and we can't find hundreds of craters 110 miles across? We're able to find a layer of iridium-laced clay no thicker than a pencil—and we can't find hundreds of holes up to 24 miles deep? We're able to find 65-million-year-old footprints less than 12 inches across; we're able to find 80-million-year-old dinosaur eggs only eight inches long—and we can't find hundreds of craters as big as three states?

Sure, a few sneaky holes may be hiding in the seas, but since 29% of our planet is above water, shouldn't 29% of those cosmic wrecking balls have fallen on land? At that rate, there should be 152 huge craters punched willy-nilly into the globe like so many finger holes in a bowling ball.

That would be enough craters to destroy every state in the United States east of the Mississippi River. That would be enough craters to wipe out France, Denmark, England, Germany, Portugal, Italy, Austria, and Spain. Our planet should be pockmarked and disfigured like the moon.

And we can't find the holes?

But impact believers never quit.

Chicxulub is far bigger than we thought, said geophysicist Virgil "Buck" Sharpton at the Lunar and Planetary Institute in Houston. Analyzing nearly 7,000 readings of gravitational strength and magnetic field data around Chicxulub, Sharpton and his colleagues came to a startling conclusion.

When that doomsday rock crashed into our planet, said Sharpton, it left a subtle bull's-eye pattern of gravity variations with a diameter of 185 miles, making Chicxulub the largest known crater in the inner solar system, bigger even than the Mead Crater on Venus. This should confirm the impact theory for all time. (*Science*, 1993)

But the debate rages on.

Who can be sure that those gravitational and magnetic anomalies were indeed caused by a meteor? Or even that Sharpton and his colleagues are reading the data properly? "Geologists are intrigued," said Richard A. Kerr in the same issue of *Science*, "but far from convinced." "It's far from certain," said William Hartmann of the Planetary Science Institute in Tucson.

Other doubters include Richard Pike of the U.S. Geological Survey, who, after studying the new analysis, is still not convinced that the subtle outer ring is real.

Mark Pilkington and Alan Hildebrand aren't convinced either. Bore holes drilled just outside the old 125-mile outer limit, said Hildebrand, reveal undisturbed rock. Not so, said Sharpton, I've seen the samples . . . "the drillers were wrong."

Which group is right? At this point, who knows? But let's look at this from a different angle.

If Sharpton is right, if Chicxulub is indeed 185 miles in diameter, you could stuff, not three, but five states into a hole that big. Delaware, Connecticut, Massachusetts, Rhode Island, and Vermont would all fit in, with room left over for Washington, D.C. and a dozen of the world's other major capitals.

We can't find hundreds of holes, each as big as five states?

A crater that big, said planetary scientist Eugene Shoemaker of Lowell Observatory in Flagstaff, "would pretty much wipe out the whole metropolitan corridor from Boston to Washington, D.C."

Now don't get me wrong. I agree emphatically, there is a cycle to mass extinctions. And yes, they're connected to our periodic trips through the galactic plane. But did meteors do it? No. Meteors have nothing whatsoever to do with extinctions. Never have, never will.

The Nemesis people have hitched their wagons to the wrong star.

Savants and priests of the earliest cultures knew that the earth was flat. To think otherwise was absurd. If the earth had another side beneath our feet, rain would have to fall upwards, water would not stay in lakes, and people would have to walk upside down.

—S. WARREN CAREY

3

. . . .

DISAPPEARING ACT

. . . .

"The arrogance of the impact-extinction people is unbelievable," bristles Robert Bakker. "The real reasons for the dinosaur extinctions have to do with temperature and sea level."

No slouch in the humility department himself, the pony-tailed Bakker, Ph.D. has been poking holes in paleontological balloons since his undergraduate days at Yale. Self-confessed heretic, one-time adjunct curator at the University of Colorado's Museum in Boulder, author/illustrator of the 1986 book *The Dinosaur Heresies*, and consultant on the movie *Jurassic Park*, Bakker is one of the world's foremost—and outspoken—dinosaur experts.

And he's right.

The enormous sea level drop at the end-Cretaceous, along with the change in weather (it turned deadly cold), provide the key, not only to the dinosaurs' demise, but to every major extinction in history.

It's one of those "Oh, by the way" things. "Oh, by the way, end-Cretaceous seas stood 1,100 to 1,800 feet higher than today." Whoa! That's a third of a mile! Wouldn't that mean that almost half of the United States and Canada was under water?

That's exactly what it means.

End-Cretaceous seas stretched north from the Gulf of Mexico to the Arctic Circle and east from the Rockies to the Appalachians. Where midwestern farmers now plant their fields of wheat and drive to their churches on Sundays, huge mosasaurs (giant sea lizards) once cavorted in waters some 300 feet above their heads, swimming in lazy circles where steeples now pierce the sky.

Real sharks plied the seas where modern-day lawyers sit and scheme in their high-rise conference rooms. Eyeball-to-eyeball with their ancient brethren, their coveted offices on the 30th floor would have been underwater. Even at 35-stories tall, today's soaring buildings would have looked short and squat, barely poking up through the water like those ubiquitous oil platforms you see dotting the Gulf of Mexico.

Then poof! The water drained away almost overnight, catching those mosasaurs and sharks without warning.

Maybe it didn't drain away.

Maybe it evaporated. Paleontologist Kenneth Hsü found huge rock salt formations more than one mile thick under the balmy waters of the Mediterranean which indicate, he believes, that the entire sea once almost totally evaporated. It evaporated over an area of 96,000 square miles, said Hsü, in his 1986 book *The Great Dying*, and to a depth of three miles.

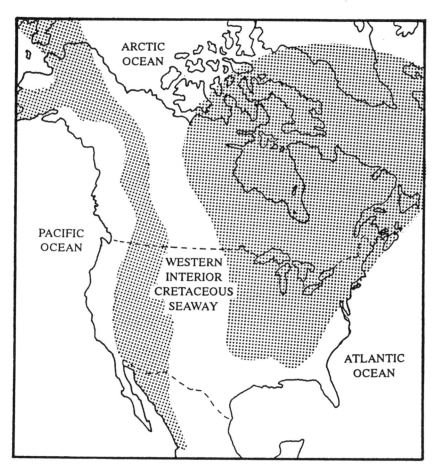

Western Interior Cretaceous Seaway. Shaded areas represent what scientists believe were two land masses separated by a vast inland sea. The rest of North America was underwater. (After John R. Horner, *Digging Dinosaurs*, 1988.)

Toss the Appalachians into water that deep and they'd sink out of sight. Then do it again, and again, because three miles deep is three times deeper than the Appalachians are tall. What could make an area bigger than all five Great Lakes combined evaporate?

Did end-Cretaceous seas evaporate too? They must have; otherwise, where did the water go? But wait. It was too dark. And too cold. With so much iridium, ash, dust, carbon, and seabed swirling through the skies, it should have been as black and as cold as the inside of a coffin. Water can't evaporate when it's cold.

Temperatures should have dropped dozens of degrees and stayed there for years. Frigid conditions may have persisted for five to ten years, said Virginia Morell in *Science*. It would have been like an ice age, or a long nuclear winter. "Even the most highly winterized dinosaur," said Morell, "would have had trouble coping with that."

That's what "should" have happened. But for some reason, water temperatures increased. Went *up*, mind you, not down, by 5° to 8°C (9° to 14°F).

Some experts, such as R. G. Bromley (1979), and Kenneth J. Hsü (1982), think sea temperatures increased an incredible 10°-12°C (18°-22°F). The increase was abrupt and short-lived.

How could temperatures in the seas have been rising while temperatures in the skies were falling? I don't think anyone appreciates what an important question this is.

I shudder when I think of how much energy it takes to raise the temperature of my backyard swimming pool a mere eight degrees—and that's just one tiny pool. You could stuff more than 22,000-billion swimming pools into the Atlantic Ocean. And it's smaller than the Pacific.

How much energy would it take to raise the temperature of 22,000-billion swimming pools by eight degrees? All at once? It's mind-boggling. Where did the energy come from?

Not from the sun, that's for sure. Solar heaters don't work too well in the dark.

Whatever caused it, evaporation, or something else, it's still a fact: A fifth of a mile of ocean disappeared. And it was not a one-time affair.

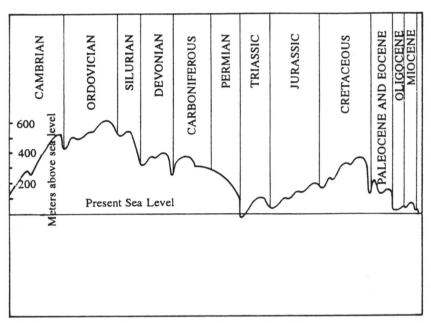

Hallam chart, variations in sea level through time. By permission,
Lapworth Professor Anthony Hallam, University of Birmingham,
School of Earth Sciences.

First noticed by Norman D. Newell some 30 years ago, sea levels
plunged by hundreds of feet at almost every extinction in history: at
the end-Cambrian, end-Permian, end-Devonian, end-Ordovician,
end-Cretaceous, end-Triassic, end-Eocene, and again (by 350 to 400
feet) at the mammoth extinction of 11,500 years ago.

*We're on the brink, I believe, of another abrupt sea level decline
right now!*

The drops are easier to picture on a graph. Anthony Hallam of the
University of Birmingham, U.K. drew a chart (above) showing sea
level changes during the past 600 million years. Dr. Peter Vail and
his colleagues at the Exxon Production Research Company drew a
chart (next page) of their own.

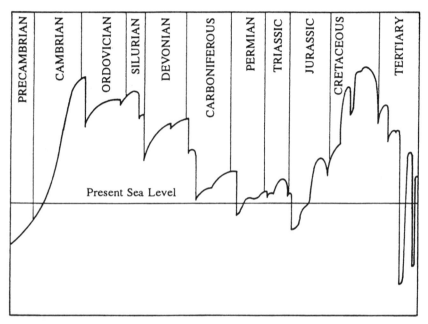

Vail chart, variations in sea level through time. By permission,
Dr. Peter R. Vail, Rice University, Dept. of Geology and Geo-
physics. Patterned after a drawing in the 1986 book *The Earth*,
Peter J. Smith, editor, Andromeda Oxford Ltd., publisher.

Though Vail and Hallam may not always agree on how high the
seas stood at any given time, or on how fast the changes may have
occurred, their disagreements, for our purposes at least, aren't all
that important.

What is important is that they essentially agree on when the waters
fell, and when the waters began rising again.

Is it just a coincidence that each sea level decline occurred at an
extinction? No. Did meteors cause every decline? No. We've got
something far more destructive on our hands here than some puny
little asteroid.

How could an asteroid, landing in a farmer's field in Iowa, heat all of the world's oceans by eight to 22 degrees? How could an asteroid, striking the Yucatán Peninsula, cause sea level declines in Denmark and pulverize their fish?

That's right, pulverize their fish! Denmark's boundary clays contain so many shattered fish fragments that they're called the fish clays (the *fiskeler*).

The fish clays, located on the coast south of Copenhagen near the town of Stevn's Klint, consist of a four-inch layer of dark-gray clay sandwiched between a layer of white chalk below and two layers of white limestone above. Except for the broken bones, the fish clays contain almost no hint of life. (For what it's worth, the fish clays contain 200 times the normal amount of iridium.)

And how could an asteroid, no matter where it crashed, kill microscopic plankton in every ocean of the world?

Gotcha! We've gotcha on that one, say impact believers. Acid rain killed the plankton. We know there was acid rain, they insist, because the lowered seas were always very acidic. Acid rain, caused by an asteroid, killed the plankton. Case closed.

One question please. Doesn't rain fall on creeks, streams, and lakes as well as on the seas?

If it was acid rain, why were freshwater animals spared?

Fourteen percent or less of freshwater dwellers died, says Charles Drake of Dartmouth College. Freshwater mollusks underwent no significant deaths at all. What kind of mad killer would destroy 70% of all sea dwellers, but only 14% of freshwater inhabitants?

It doesn't make sense. In today's world, lake dwellers die from acid rain long before ocean dwellers do. Forget acid rain. We've been accusing an innocent bystander.

All sorts of weird things happened to the K-T seas. Not only acidic, they were also anoxic (deficient in oxygen).

Scientists assume they were anoxic because of the black muds that settled to the bottom of the seas; black muds which later became black shales; black shales resting on top of apparently healthy reefs. "The sea suffered from indigestion," said Alfred Fischer of Princeton

University. So much carbon and soot poured into the seas that they couldn't metabolize it all.

Again, it's a recurring thing. Black shales are found at the ends of several geologic periods, including the Cambrian, Permian, Silurian, Ordovician, Devonian, and Late Triassic. Digby McLaren at the University of Ottawa tells of black shales at the Frasnian-Famennian extinction (middle Upper Devonian), while Hallam tells of black shales at the end-Pleinsbachian, end-Cenomanian, and end-Aptian.

But why? It's a throw-up-your-hands-in-despair kind of thing. If anoxic conditions had occurred during times of low sea levels only, scientists might be able to figure it out. But widespread anoxic conditions also correlate well with high sea level stands.

Funny stuff happened in the seas with carbon, too. Marine sediments immediately following the K-T are always deficient in carbon-13 atoms. Ten times more deficient. And the deficiency occurred abruptly. The seas must have been sterile, say scientists; they contained no gradient of carbon isotopes (no change from top to bottom), just as in a nuclear holocaust.

There's more. The kinds of plankton that secreted calcium carbonate skeletons were devastated, while those that secreted silica escaped almost unscathed. More than 70% of all swimming organisms died, but only 50% of benthic organisms (bottom dwellers) in the same seabed were killed. What made bottom dwellers so lucky?

Those lowered seas were full of surprises. In direct disobedience of all laws of nature, they didn't always rise and fall in concert. Sea levels rose in some places (or so it would seem), while remaining the same in others. That's impossible of course.

It's "a matter of some controversy," says paleobiologist Steven M. Stanley of Johns Hopkins University. Take the end-Triassic, when the seas fell relative to the land in parts of Europe while rising in most of the rest of the world, including western North America. Or look at the mid-Jurassic, when sea levels fell in the North Sea while remaining at their pre-existing depths in the rest of the world.

What's going on? I mean, we learned it in the third grade. Water seeks its own level, right? And all oceans are interconnected, right?

So if the seas rise in one part of the world, they should rise all over the globe, right?

That's not what the record shows.

During the Aptian-Albian transition, the seas retreated from Australia but didn't fall in the rest of the world until Campanian times. A similar disparity occurred in the Oligocene, when sea levels fell substantially in the North Sea and northwest Africa but remained almost unchanged in Australia.

Hey, c'mon guys. How should we picture this phenomenon? Did cliffs of water surround Australia like 180-story waterfalls? Were the ancient mariners right? If they'd kept sailing west, would their boats have fallen off the edge of the world?

Or should we picture Australia as being in some quaint little valley out there in the middle of the ocean? Surfers would have loved it. Downhill all the way.

You can't defy the laws of gravity like that. Can you? You can't stack water like pancakes. Can you?

Yes, you can. Freeze it. Then you can stack it. And that's the secret. Many geological periods had ice ages, and that, many scientists insist, is where the water went. It turned to ice and got locked up on land. Same at the end-Cretaceous. Though scant evidence of glaciation has been found at the K-T, it is possible, say scientists, that a short-term period of ice and snow might not show up in the geologic record.

But even if they could find evidence of glaciation, they still couldn't explain it. How do you form continent-wide ice fields abruptly, with water that's eight—perhaps as much as 22—degrees warmer than normal?

So, are we back to a drought? Is it possible that skies were sunny and bright, and that a drought came along to dry up the seas? No, it couldn't have been sunny. Those layers of iridium, ash, carbon, and soot cannot be denied. K-T skies had to have been black.

Besides, it's impossible to have a world-wide drought. Think about it. If today's seas should evaporate, where would the water go? It would rise into the sky to become clouds. Then it would rain.

True, the rain might not fall where it came from, but regardless, like a salmon going home to spawn, its urgent yen for the sea would send it racing down every river and stream and back to the ocean.

It's a never-ending cycle. The more it evaporates, the more it rains. You might get a few localized droughts here and there, but there's no way that a fifth of a mile of ocean would stay in the sky around the entire globe for very long at all. You'd have world-wide humidity, not drought. There is simply no way to have a world-wide drought-causing drop in the seas.

It's an impossible riddle. A loss of sea water is "unlikely," say scientists. And yet . . . it seems to have disappeared.

Where did the water go?

Blame magnetic reversals.

It would be an irony if one of the very elements that led Alvarez to the planetesimal cause of the extinctions was itself the agent of those extinctions.

—MICHAEL ALLABY & JAMES LOVELOCK

4

. . . .

THE IRONY

. . . .

Thirty precious years ago, Robert Uffen warned us about magnetic reversals; thirty precious years, during which we could have been preparing for the coming disaster. Now we need to play catch-up, and we need to do it fast.

Our magnetic field (the magnetosphere) shields us from the charged particles in the solar winds, said Uffen, of Canada's University of Western Ontario. During a reversal, cosmic radiation would bombard our planet, leading to mutation or death.

I think Uffen was right.

So do many others. "Faunal [animal] changes," said Allan Cox of Stanford University, "occurred near several reversals."

New kinds of animals appear in the record "virtually simultaneously" with reversals, said James P. Kennett and N. D. Watkins of the University of Rhode Island.

"Reversals strongly influence population trends," said C. J. Waddington at the University of Minnesota.

"There is mounting evidence that a correlation does exist between major faunal extinction and geomagnetic polarity reversals," said G. C. Reid of NOAA.

During the last 4.5 million years, said James D. Hays of Lamont-Doherty Geological Observatory (now Lamont-Doherty Earth Observatory), at least six out of nine radiolarian extinctions occurred at magnetic reversals. In the equatorial Pacific, said Hays, seven out of eight extinctions took place near reversals. "The close correlation suggests a causal connection."

But why? It all goes back to radioactivity. "We have long known," said paleontologist John F. Simpson, "that exposure to ionizing radiation increases mutation rates." During a reversal, the earth's magnetic field strength would drop to zero, thereby allowing excess radiation into our skies. Magnetic field strength, Simpson noted, is decreasing right now.

Let me repeat. Magnetic field strength is decreasing right now!

But others disagreed. It would make no difference, they said, if magnetic field strength dropped to zero. Even if a charged particle weren't deflected by the field, it would need to be traveling at an almost impossibly high rate of speed to make it through our atmosphere and to the ground. (Some cosmic particles do travel at impossibly high rates of speed. More later.)

Others think a weakened magnetic field might reduce the ozone layer. And the ozone layer, they say, like the magnetosphere, protects us from deadly outside forces. Lose it, and ionizing radiation would come crashing into our skies.[1] This would lead to the forma-

---

1. Radioactive elements such as carbon-14 are constantly created high in our skies when the speeding particles in cosmic rays collide with nitrogen atoms in the atmosphere. This process, called ionization, adds or removes electrons from atoms or molecules which were previously neutral. Add a neu-

tion of large quantities of nitric oxide, which catalytically destroys ozone. (Crutzen *et al.*, 1975)

One thing upon which most scientists agree is that the earth's magnetic field was reversed at the end-Cretaceous. It's hard to deny, given that the last dinosaur remains are always found in reversely magnetized sediments. The period of time during which it remained reversed, about 500,000 years, is called C-29-R, which stands for Chron Cenozoic 29-reversed. (A polarity epoch is called a chron.)

Remember, that's why Alvarez went to Gubbio in the first place, to study reversed polarity. But he got sidetracked by that seductive layer of iridium and came back with the asteroid theory instead.

What an irony!

Peel back the layers of mystery surrounding polarity reversals, peel back the layers of clay at Gubbio, and you'll solve the entire extinction enigma.

Let's begin peeling. How do we know that our magnetic field was reversed at the end-Cretaceous? Through magnetostratigraphy, the study of the magnetic properties of ancient layers of sediment (strata) now hardened into rock.

Magnetic materials, such as magnetite, occur in all rocks. Like miniature magnets, they're tiny pieces of ferrous metal that aligned with the earth's magnetic field as the rocks were being formed. As sedimentary grains drift to the bottom, the earth's magnetic field twists them in the water (like compass needles) until they align with the field. By determining their lie (which way they point), scientists can tell which way was north when the strata formed.

So too with igneous rocks and basalt. Magnetites in magma and lava also align with the field. Non-magnetic while hot, their iron and titanium oxides become magnetic as they harden and cool through the Curie temperature, thus writing a record in stone as to when the

---

tron to almost any atom, said Willard F. Libby, one-time commissioner of the Atomic Energy Commission, and it will become heavier, and frequently radioactive. Beryllium-10 (^{10}Be) is another radioactive element created in the sky. So is helium-3. So is tritium (radiogenic hydrogen). Small amounts of radioactivity are falling on your head this very second.

rocks were created. (The Curie temperature ranges from 200° to 680°C, depending on what is being heated.) To paleomagnetists, lava is a giant recorder storing information from the past much as a computer stores data in its magnetic memory.

Magnetites not only orient themselves parallel to the existing magnetic field, they also tilt. Hold a bar magnet above a larger one, and it will point downward toward the end of the larger magnet. Compasses would tilt downward too, but most have been doctored to eliminate the tilt. Hold an unaltered compass above magnetic north and it too will point straight down, straight south.

When magnetic north is in the Arctic as it is today, the north ends of magnetite tilt downward in the northern hemisphere and upward in the southern hemisphere. This tilt, sometimes called magnetic inclination, other times the angle of the dip, also shows which way was north when the rock was created.

That's how reversed polarity was discovered in the first place. Bernard Brunhes, a French geophysicist, found "wrong way rocks" in volcanic basalt in the plateau country of central France in 1909. At the time, no one understood their importance.

In 1928, Motonari Matuyama, a professor at the Kyoto Imperial University, shocked the scientific world. Conducting experiments on volcanic rocks in Japan, Matuyama made the daring proposition that our magnetic field had "flipped" sometime during the past few million years.

Then other scientists found reversed magnetism (called soft magnetism) in small portions of lava flows struck by lightning. The issue got cloudy. Does reversed magnetization occur because of a worldwide polarity flip, or does something merely happen now and then to the mineralogy of individual rocks?[1] If it were a complete reversal, shouldn't all rocks around the world during any given time period point toward the same spot? But how could anyone know with certainty when the rocks were formed?

---

1. Today, we know that soft magnetism occurs only in isolated instances. Sometimes it's caused by lightning, other times by the chemical make-up of a particular rock.

It took half a lifetime for science to catch up with Matuyama's inspired guess. In 1963, refinements in potassium-argon dating finally gave him the nod.

Since volcanic rocks contain radioactive potassium, their ages can be confirmed by radiometric dating. Their argon doesn't begin accumulating until they cool below the molten state, in essence giving them an atomic clock that begins ticking when their magnetism is acquired. Using reversals, scientists can now read the history of the ages with much greater accuracy.

Periods of reversed or normal polarity are divided into magneto-stratigraphic epochs. To honor the work of the early pioneers, today's epoch, which began about 780,000 years ago, is called the Brunhes Epoch. The epoch before today's, an epoch of reversed polarity, is named for Matuyama.

In addition to full-scale reversals, our magnetic field is sometimes fickle. It moves away from north for short periods, then moves back. These movements, called magnetic excursions, are found in lava flows in many parts of the world from many different periods.

Excursions usually began suddenly as the North Pole moved rapidly and smoothly toward the equator. Sometimes it popped back to its original position almost immediately. Other times it crossed the equator and moved part way through the opposite hemisphere before swinging back to its near axial north-south position. Excursions, many paleomagnetists believe, were aborted reversals.

There's still a lot to learn about magnetic reversals, but one of the most exciting discoveries is that they recur in a pattern. Reversals return like clockwork, said David M. Raup of the University of Chicago, about every 30 million years. And they do it in phase with the extinction cycle.

Don't let that one slip by you, it's deadly important.

The reversal cycle matches the extinction cycle!

Why do reversals occur in cycles? Two Indian geophysicists, J. G. Negi and R. K. Tiwari, think they know the secret. The pattern corresponds remarkably well, they said, with the solar system's cyclic vertical oscillations through the galactic plane. It also correlates well with the cosmic year (the time it takes our solar system to orbit the galaxy). Since the frequency of reversals seems to peak about every 285 million years, said Negi, our magnetic field may be controlled by our galactic motion.

Craig Hatfield and Mark Camp at the University of Toledo in Ohio came to a similar conclusion in their paper "Mass Extinctions Correlated with Periodic Galactic Events." So did Johann Steiner at the University of Alberta.

A funny thing about the correlation, said Steiner, is that the reversal cycle also correlates with periods of crustal deformation . . . and with periods of glaciation. Major glaciations have occurred twice per revolution, about every 140 million years. The driving engine, said Steiner, may be extraterrestrial. "The source of energy may have to be looked for in the dynamics of the Milky Way galaxy."

Major periods of glaciation, and of crustal deformation, match the magnetic reversal cycle!

What causes magnetic reversals, anyway? Remember playing with those Scotty Dog magnets when you were a kid, trying to put their noses (their north poles) together? They'd spin around so fast that you could barely see them move. If the earth has a giant magnet inside it like those Scotty Dogs, could our entire planet have somehow flipped upside down?

No, no, and no again. There is no evidence, scientists insist, that such a thing ever happened, or ever could. Besides, there couldn't be a solid magnet inside the earth anyway; it's too hot. The earth's center, they remind us, is made of boiling liquid rock. And, liquid or solid, all materials lose their magnetism when heated past the Curie temperature.

So. Why do reversals happen?

No one knows. For that matter, no one knows why we have a magnetic field in the first place. After years—no, after centuries—of intense scientific research and debate, our magnetic field is one of the best described and least understood wonders of the universe. "There is an embarrassing lack of theory," said Allan Cox, "to account for the present geomagnetic field."

Here's what we *think* we know. Our magnetic field is electromagnetic. As our planet rotates, magnetism is induced in much the same way that it's induced by the flow of an electric current through a coil of wire. The earth's core must be like a giant dynamo.

That doesn't seem possible, does it? It doesn't feel as if we're turning fast enough to create an electric current. Indeed, as we sit in our easy chairs on the surface of the earth, it doesn't feel as if we're moving at all.

But our senses deceive us. Look at the numbers. It's about 24,000 miles around the earth at the equator. Since the earth makes a complete revolution every 24 hours, that means we travel 24,000 miles every 24 hours. Divide 24 hours into 24,000 miles, and you'll come to the shocking realization that we and our easy chairs are whirling around the global axis at about 1,000 miles an hour (at the equator). That kind of speed, in theory at least, creates the dynamo effect.

But we're not talking theory here, we're talking fact. Polarity reversals do occur. And we're back to the same old question. Why? All speculation eventually leads to the galaxy. There must be an unknown galactic force, scientists agree, that causes reversals.

It would be easiest to blame it on a stronger magnet lurking somewhere in the sky. A stronger magnet provides a couple (some call it a torque) which tries to turn the other magnet around. If the weaker magnet refuses to turn, then its magnetic field reverses. But we know of no galactic force that could exert that kind of pressure on the earth.

What about an asteroid? Yes! Yes! Yes! shout impact-believers. If a large celestial body were to enter our atmosphere at an extremely high rate of speed, and if it were spinning very rapidly, it might be surrounded by a spinning cloud of ionized particles. A super-duper dynamo, so to speak, it might somehow alter the earth's polarity. Their theory hasn't attracted many adherents. In real life, it seems, asteroids do not spin.

Another theory suggests that the impact of a large body striking the globe—gee, an asteroid, maybe?—could cause major agitations in the earths's core, leading to a reversal. This is a credible theory. Hammer almost any magnet hard enough and it will lose its magnetization. Meteors would therefore be the cause of extinctions, and reversals proof of impact.

Yet another theory involves explosions. If a giant supernova with huge explosions were to occur nearby, the blasts might create a magnetic shock wave strong enough to cause the flip. Great idea, except for one minor detail: There are no heavenly bodies close enough to earth to have done it.

So what's the answer? Like the extinction theories themselves, many reversal theories sound quite plausible. But only one theory can explain why reversals keep coming back like clockwork: Our periodic vertical trips through the galactic plane are more important than we realize.

Some scientists believe the galactic magnetic field may be oriented in opposite directions above and below the galactic plane. When we pop through the plane our magnetic field reverses in order to remain in sync with the galaxy (Morris and Berge).

Others speculate that the galactic magnetic field may be stronger the closer we get to the galactic equator; that its most intense lines of force lie in the plane. When we bounce through the plane, those intense lines of force dump excess cosmic rays on our heads (Hatfield and Camp).

The galactic magnetic field also appears to be stronger as we go through its spiral arms, said W. H. McCrea of the University of Sussex, United Kingdom. We're just inside the inner edge of an arm right now, said McCrea, inside the Orion Arm.

And that brings us to the alignment theory. Like the earth, our galaxy has a south pole (the south galactic pole) and a north pole (the north galactic pole). It also has a magnetic equator (the galactic equator, or galactic plane). The magnetic equator is shaped like a flat circular disk some 100,000 light-years (30 kiloparsec) across. All stars lie in or very near the disk. Our solar system lies about three-fifths of the way from the galactic center (in the constellation Sagittarius), about 32,600 light-years, or 10 kpc, away. (One kpc equals about 3,258 light-years. One light-year, the distance light travels in a year at the speed of 186,234 miles a second, is about 5.88 trillion miles, a shade under six thousand billion miles.)

Our solar system also has a north and south pole; the north celestial pole and the south celestial pole, and a celestial equator (or celestial plane). Just as all stars lie within the galactic plane, so do all planets lie within the celestial plane, never deviating by much more than 1° from their common plane. This is hardly coincidental.

Our entire solar system revolves clockwise around the galactic center. And it revolves *fast,* speeding through space at 155 to 200 miles per second. Within the solar system, of course, is the earth, which also revolves.

What we have here, then, are three spinning dynamos: the earth, the solar system, and the galaxy, one within the other within the other. Three revolving dynamos, with their equators lying at different angles to each other. And as long as they remain that way, everything stays in balance.

But every 28 million years, when we pop through the galactic plane and our magnetic lines of force come into alignment, the trouble begins. It's that alignment, I submit, that causes the reversal.

When the lines of force in the solar system line up with the lines of force in the galaxy, the entire solar system's magnetic field, along with the earth's, reverses. As it reverses, magnetic intensity decreases, and cosmic rays crash into our skies.

This is the stronger magnet we've been looking for, the one that provides the torque. (It may not be stronger, but it sure is bigger.) Since it's unlikely that our insignificant little solar system could force the entire galactic magnetic field to reverse, we're compelled to make the flip. Then wham! Radioactivity rains onto our planet.

And that's *one* of the ways by which magnetic reversals cause mass extinctions.

There are well-confirmed ob-servations of a correlation between magnetic reversals and extinctions.

—Ian Crain

5

. . . .

RULER OF THE UNIVERSE

. . . .

Magnetic forces rule the world. Sometimes ruling with benevolence, other times with an iron fist, their powers reach into every corner of the universe. Ignore magnetic forces, and every motor in the world will stop running. Ignore magnetic forces, and your airplane will fall from the sky. Ignore magnetic forces, and your computer will crash and burn out.

Ignore magnetic reversals . . . and you will die.

To see how powerful a magnetic reversal can be, look at the sun. The sun, as you may already know, is a giant unshielded nuclear reactor made of millions upon millions of never-ending nuclear explosions. Nuclear explosions are the very engine that drives it.

But what you probably do not know is what causes those explosions in the first place. Magnetic forces are the cause, the very trigger, of millions of nuclear explosions in the sun. Magnetic forces are the starter and the fuel that keeps them going.

The explosions occur when hydrogen nuclei, or protons, in the sun fuse into helium nuclei. The protons whose fusion is required to start the reactions, said Robert W. Noyes, in his book *The Sun, Our Star*, are mutually repelled, both by their electrical charge and by nuclear repulsive forces. Magnetic forces somehow defeat that repulsion.

Our sun is a magnetic star. Like the earth, the sun has a north (solar) pole, a south (solar) pole, and an equator. And, like the earth, the sun rotates. Its equatorial regions rotate slower than the poles, in a process called differential rotation.

It's hard to believe that something as big as the sun (you could stuff a million earths into it) could rotate very fast. But even with its huge diameter of 840,000 miles, the sun makes a complete revolution every 27 days.

That speed, more than 4,000 miles an hour at the surface, creates millions of magnetic fields all twisted together like a huge ball of burning twine; millions of intertwined strands of intense, pulsating magnetic forces that heat the sun's corona to more than one million degrees. *One million degrees—all from magnetic forces!*

There's a grid system, not totally understood, which contains the explosions in groups, and prevents them from crossing into each other's territories; a grid system with its edges defined by yet other powerful magnetic lines of force.

About 900 miles square, each grid, or granule, is in a continual turbulent state of eruption. These Texas-sized granules cover the entire sun. Looking at the sun through a telescope at Kitt Peak near Tucson, Arizona, observers describe it as "a tightly packed ball of churning giant grains of wheat," or "a pot full of boiling oatmeal."

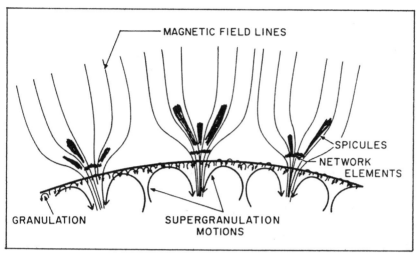

Schematic cross-section of the sun's surface. Large-scale convective cells—supergranulation—sweep magnetic fields to their outer edges where the field collects and strengthens. The spacing of the pattern is about 30,000 kilometers, or two and a half times the size of the earth. (Reprinted by permission of the publishers, from *The Sun, Our Star* by Robert W. Noyes, Cambridge, Mass.: Harvard University Press, Copyright © 1982 by the President and Fellows of Harvard College.)

Like so many fiery bubbles in a witch's cauldron, the granules constantly move, heave, bubble, and turn over. During each eruption, matter is flung upward for about 60 miles at about 1,000 miles per hour. Then it cools, drains back down, and sinks 60 miles into the sun. Round trip, a granular eruption takes about eight minutes. Then matter is flung upward again. Granulation takes place 24 hours a day, day in and day out.

Then there's supergranulation, where bunches of granules join together into bigger bubbles, working as a unit in a larger churning process about 18,000 miles across. (Picture thousands of beach balls bunched together, all turning to their own special tune. That's supergranulation. Now see each of those beach balls filled with hundreds of ping-pong balls, with each ping-pong ball revolving, too. That's "regular" granulation—except the ping-pong balls and the beach balls are made of fire.)

Gigantic eruptions, called flares and prominences, continually burst out of the areas of granulation. Like giant fingers of fire leaping high into the sky, they form fiery loops many times the size of the earth and greatly increase the already steady flow of particles in the solar winds. Magnetic forces fling them upward against a gravitational field 40 times that of the earth. (That's no small achievement. The mass of a good-sized prominence, said Noyes, about 10^{16} gram, equals that of a small mountain.)

The most violent of all solar events, there's so much energy in a flare that one flare—just one—could supply the energy needs of the entire United States for almost one million years. There's as much energy in a flare, said Noyes, as you'd obtain from exploding two and a half billion one-megaton hydrogen bombs.

Like giant lightning strokes, flares can throw arcs of blazing force across distances the width of Texas in seconds, and have been known to shoot 36,000 miles across the face of the sun in less than five minutes. At that speed, more than 430,000 miles an hour, if they exploded on earth they'd traverse the United States east to west in 15 seconds, and encircle the globe in about three minutes.

Then they just sit there like massive bolts of lightning crackling and popping in the sky. But unlike lightning, which disappears within a few seconds, these blazing arcs of force can survive for hours or days. Some last for weeks, others for months. Magnetic fields apparently support those loops of fire, said Noyes. They're "the invisible ropes that hold them up."

For some reason, prominences and flares don't travel sideways. The sun's magnetic fields act as barriers which prevent their movement, perhaps the same kind of barriers that hold the granulation process in check.

Flares and prominences can shoot to staggering heights. During a flare on June 1, 1991, observers spied a bright jet of fire probing more than 400,000 miles (more than 50 earth diameters) into the sky above the sun. Then it hung there for more than six hours. "By the next day," said a story in the *Seattle Times*, "a speeding cloud of particles, blasted free of the sun's gravity in the eruption, began squashing the earth's magnetic field . . . [and] northern lights lit up the skies as far south as Denver and Northern California." Auroras *always* occur right after solar flares, and are *always* associated with geomagnetic disturbances.

Did you know that?

Did you know that magnetic forces 93 million miles away (the distance from the earth to the sun) are the switch that turns on the northern lights?

Then come sunspots, those mysterious dark spots sometimes visible to the naked eye, which revolve east to west around the surface of the sun, the same direction that all planets move.

Ancient Chinese, Japanese, and Korean astronomers were aware of sunspots, and the Greek Theophratus, a pupil of Aristotle, reported them in 300 B.C. But they didn't come under much scrutiny until 1610 when reported by Galileo, the Italian scientist and inventor of the telescope. No longer quite so mysterious, we now realize that sunspots are areas of extreme magnetic turbulence.

Sunspots are like oscillating magnetic dynamos. Their most predominant property is their intense magnetic field. It's the root cause of a sunspot's very existence.

Though typically about 20,000 miles in diameter, sunspots can be as small as the state of Arizona (a few hundred miles across), or they can be huge. One such giant, measuring 80,000 miles across, thousands of times bigger than the earth, erupted on May 9, 1991.

The larger the sunspot, the stronger the magnetic field. The sun's normal magnetic field strength is about the same as the earth's, less than one gauss. As opposed to the sun itself, a sunspot's magnetic strength is immense, ranging from 500 to 4,000 gauss, 10,000 times more powerful than the earth's.

Sunspots burst through the sun's surface every 11 years in gigantic nuclear explosions. And that's the point of this chapter. Those nuclear explosions are a direct result of a reversing magnetic field.

Ring all the bells and blow all the whistles. This is the proof I promised! Those gigantic nuclear explosions in the sunspots are the direct result of a reversing magnetic field!

At the beginning of each cycle, magnetic polarity on the sunspot reverses and magnetic north becomes magnetic south. No one knows why. (Though the spots appear about every 11 years, there's a 22-year cycle involved. That's how long it takes the polarity to reverse, then go back to its original state.)

Maybe we don't understand, but understand or not, the simple fact still remains: Magnetic reversals cause massive nuclear explosions on the sun. Aren't we part of the solar system? Shouldn't we be subject to the same laws of nature? If magnetic reversals on the sun cause nuclear explosions, shouldn't magnetic reversals on earth cause them too? Of course they should.

Blasting into our skies at the speed of light, galactic-sized bursts of power reach earth almost immediately after a sunspot reversal, sometimes within nine minutes. Then they begin squeezing between the earth's protective magnetic lines of force.

The earth's magnetic lines of force hurtle out of the magnetic North Pole, take a long curving loop through space around the planet to completely surround the earth, then dive back in at the magnetic South Pole. Magnetic lines of force, whether belonging to the earth or to a common magnet, always form continuous closed loops.

The solar wind squeezes the earth's magnetic lines of force on the upwind side, and extends them on the downwind side. Like a gigantic magnetic teardrop, the lines extend more than 40,000 miles into space on the daytime side of the earth (the side facing the sun) and more than 150 million miles on the nighttime side. Attracted to them, then repelled when they get too close, the energized particles speeding in from the sun constantly corkscrew around the lines.

But the lines bunch together so closely where they come out of the North Pole that the electrified particles have no room to continue their spiraling paths; no room to maneuver. This causes a few to leak out of the magnetosphere. Once released, like migrating geese headed north, they race helter-skelter into the earth's upper atmosphere where they precipitate toward the North and South Poles.

Upon reaching the upper atmosphere, the energized particles collide with its atoms and molecules, releasing radiation—much like that in neon lights—to create those undulating, waving, fluttering curtains of light that we call the aurora borealis. "Think of the northern lights as a gigantic neon sign with a high-volume electrical discharge," said Dr. Syun-Ichi Akasofu, director of the Geophysical Institute at the University of Alaska, Fairbanks.

Auroras (from the Latin word for dawn) occur primarily in the far north or far south near the geomagnetic poles. (Aurora borealis: dawn of the north; aurora australis: dawn of the south.)

This process of escaping and colliding energized particles, this atomic bombardment (they're minor nuclear collisions), goes on all the time, day and night. When an accelerated proton collides with an atom they merge, forcing the atom to split into two new nuclei. Usually called ionization, this barrage from the cosmos is also called hydrogen impact ionization, or ionization by collision.

Whatever it's called, it unlocks the energy trapped in the charged particles and instantly converts it to matter. The same sort of thing happens in tests when an electric current is shot through gases, liquids, or solids.

But a flare greatly increases that stream of particles, squashing our magnetic field even more. As the field gets squashed, the lines push further away from each other at the poles, creating "holes." This allows more particles into our atmosphere, ever closer to the equator, which in turn causes major increases in ionization as close as 47 miles above the ground, and causes the northern lights. Though they can occur as high as 600 miles above the earth, and as low as 47, auroras usually occur about 60 miles up.

Isn't that kind of spooky, to realize that magnetic reversals 93 million miles away cause nuclear reactions just 47 miles above your head? Isn't that kind of spooky, to realize that you live just 47 miles from an uncontrollable nuclear reactor?

That's during *normal* times!

The speeding particles, called solar cosmic rays, have energies as high as those in galactic cosmic rays. Although they contain some electrons, cosmic rays consist mainly (81%) of protons. Protons are positively charged nuclei of hydrogen atoms.

Some mechanism, so far not understood, preferentially accelerates these atomic particles to extremely high speeds. Accelerated to almost the speed of light (that should qualify as an almost "impossibly high rate of speed," don't you think?), some of them carry up to 10 billion electron volts. Somehow, scientists agree, the sun's reversing magnetic fields must play a major role.

Think about that for a minute! Polarity reversals play a major role in generating up to 10 billion electron volts. Polarity reversals play a major role in accelerating atomic particles to the speed of light, the kind of speed that man-made nuclear particle accelerators can only dream about. And we don't think they're important?

It's the same power that we use to create matter. Man-made particle accelerators, such as at the Stanford Linear Accelerator Center, create new matter almost routinely with headlong collisions between electrons and anti-electrons. Two miles long, the Stanford accelerator creates new electrons, *swarms* of new electrons, 40,000 times heavier than when they started. (It's a good thing electrons aren't people, isn't it? Stuff one person into an accelerator like that, and out of the other end would pop a city of 40,000.)

If a man-made accelerator only two miles long can create swarms of new electrons 40,000 times heavier than when they started, imagine what a cosmic-sized accelerator could do. Ionization rates would increase by the billions.

And if a magnetic reversal 93 million miles away can cause miniature nuclear explosions just 47 miles above our heads and create new matter, doesn't it seem possible that a reversal right here on earth, just zero miles away, could send waves of nuclear explosions cascading around the globe, dumping untold amounts of radiation on our heads? I think it does.

Magnetic reversals on the sun do a lot more than paint pretty neon pictures in the sky. As sunspot activity waxes and wanes, so does the production of nitric oxide. So does the production of radioactive carbon-14. (Carbon-14 production increases during sunspot minimums.) In a direct example of energy turning to matter, part of the newly created carbon-14 enters the life-chain of both plants and animals. Trees grow taller and thicker with their growth rings further apart. The change in thickness occurs in exact synchronization with the sunspot cycle.

Sunspots affect the earth in other ways, too. During solar flares and for two to six hours after, our magnetic field suddenly increases 20 to 30 gammas, causing massive magnetic storms. The storms disrupt electronic communications and cause fade-outs in shortwave-radio emissions. In power lines, fluctuating magnetic fields caused by sunspots can produce large electrical currents. Sometimes, said Patrick Huyghe, in *Science Digest*, erupting flares cause electrical surges in the ground itself. *In the ground itself!*

Overloading transmission and telephone lines to the breaking point, sunspot-caused power surges can burn out the components in power stations. A flare in 1989, for example, virtually shut down the entire Canadian power network.[1] And oil companies are painfully aware of what sunspot reversals can do. Flare-induced electrical currents at high geomagnetic latitudes have been known to cause corrosion in metal pipelines.

Fluctuating rapidly during a flare, the earth's magnetic field produces bizarre effects. Pilots flying near the poles sometimes report erratic compass readings. And pity the poor homing pigeon, who seems to navigate by an inner magnetic sensor. During a normal homing pigeon race, 70% to 80% of the contestants make it home. But when a major flare erupted during a recent 500-mile race, only four percent made it back to the roost.

Even lightning is influenced by sunspots. Studying lightning strikes on power distribution systems in Britain, scientists discovered an unmistakable 11-year periodicity in phase with the sunspot cycle. Maximum thunderstorm activity also correlated with the cycle.

So too with volcanoes. Volcanic activity tracks the sunspot cycle "passably well," some scientists insist. (Noyes disagrees: The case that sunspots cause volcanoes, says Noyes, "is almost as implausible

1. From Douglas, John, *et al*, "A Storm From the Sun," *EPRI Journal* (Electrical Power Research Institute), Jul/Aug 1989.

as the case that volcanoes cause sunspots!") But how will he explain
the recent eruption of Mount Pinatubo in the Philippines? Or Mount
Unzen in Japan? Or Mount St. Helens? Or all of the other eruptions
in tune with the sunspot cycle?

Even earthquakes! Pointing to large tremors in 1857, 1881, 1901,
1922, 1934, and 1966, some scientists go so far as to say there's a
22-year regularity to earthquake activity which matches the sunspot
cycle "remarkably well."

It doesn't exactly look like a 22-year cycle does it? But keep your
eye on the ball as they move the numbers around. Since the quake
of 1922 was much weaker than the others, maybe it didn't release all
of its energy, which made the 1934 quake occur ten years before its
time—and presto!—a 22-year earthquake cycle.

Sounds like a lot of number massaging to me. But the October
1989 quake in San Francisco does plug into the average rather neatly. Perhaps we should keep an open mind.

Listen to astrophysicist John Gribben and physicist Stephen
Plagemann in their book *The Jupiter Effect*. "There is a strong case,"
they said, "that unusual solar activity causes increased seismic activity." "Small-scale seismic activity may be enhanced for weeks after
a flare." "Of the eight earthquakes centered within a radius of 48
miles from the entrance to San Francisco Bay which have occurred
since 1836, every one of the eight has occurred within 2 years of a
sunspot maximum."

Even our weather is affected by sunspots.

"There is a large body of evidence," said J. Murray Mitchell of
the Environmental Data Service in Silver Spring, Maryland, "that
climate varies with either the 11-year sunspot cycle or the 22-year
'double' sunspot cycle."

"A cause and effect relationship links changes of the Earth's
magnetic field and climate," agreed Goesta Wollin of Lamont-Doherty Earth Observatory (with colleagues Ericson and Ryan), "and both
effects may be the result of varying solar activity."

Even glaciation!

Alaska's ice masses advance and retreat in direct correlation with
the sunspot cycle, said glaciologist Dr. Maynard Miller. The growth

and decay of the ice fields, said Miller, director of the Juneau Ice-field Research Program, is caused by cyclic solar storms.

One surprising discovery is that sunspots are colder than the rest of the sun by several thousand degrees. Colder. Not hotter. Soon after the sunspot's magnetic field appears, the area gets cooler than the rest of the sun, and the flaring stops.

At the same time, the spectrum lines of many elements get stronger. Spectrum lines of carbon monoxide, calcium, silicon, fluorine hydrides, titanium and zirconium oxides, along with hydrides of nitrogen, carbon and magnesium, all show a much greater strength in sunspots. Such a strengthening, scientists believe, indicates an increased abundance of these elements.

This cooling has not been explained, but I think we're witnessing Einstein's theory at work. Instead of burning matter to create heat, sunspots consume heat to create matter. One thing is certain however, the cooling is caused by the presence of an extremely powerful magnetic field.

Regardless of what causes the cooling, magnetic reversals in sunspots produce profound effects on the earth. And here they are again in a nutshell:

Magnetic reversals 93 million miles away create large electrical currents on earth. Magnetic reversals 93 million miles away cause power surges, communication disturbances, power station burnouts, and turn on the northern lights. Magnetic reversals 93 million miles away affect thunderstorms, affect our weather, affect our glaciers, and may cause earthquakes and volcanoes.

Magnetic reversals 93 million miles away send huge tongues of blazing force soaring 400,000 miles into the sky above the sun carrying the power of two and a half billion hydrogen bombs. Magnetic reversals 93 million miles away cause nuclear reactions just 47 miles above our heads.

What would happen if a magnetic reversal should occur right here on earth, just zero miles away?

I think we're about to find out.

*There could be an intriguing
possibility of a direct link
between large-scale tectonic
processes such as ocean-floor
spreading and geomagnetic reversals at the Earths's core.*

—ALLAN COX

6

. . . .

CONTINENTAL DRIFT

. . . .

What would happen if a magnetic reversal occurred right here? The
same things that happened in the past. Earthquakes, floods, volcanoes, giant snowstorms, rising land, plummeting sea levels—you
name it—tectonic activity would go bonkers.

It all goes back to Alfred Wegener and his theory of continental
drift. Wegener, a German meteorologist, came up with his theory in

1912 after noticing that Brazil, on the east coast of South America, would fit neatly into the underbelly of Africa.

They were once connected, said Wegener, in his iconoclastic book *Die Entstehung der Kontinente und Ozeane* (On the Origin of Continents and Oceans), as a single chunk of land called Pangaea.[1] Now they've pulled apart and float across the face of the globe like wandering rafts on the sea. Look at a globe. The two continents would fit together like pieces of a giant jigsaw puzzle.

"A ridiculous manifestation of unprincipled and unfettered imagination," sneered scientists. "*Ein Märchen.*" "A pipe dream, a beautiful fairy story." "Utter, damned rot," quoth the president of the prestigious American Philosophical Society in Philadelphia.

But when Wegener's books were translated into English in the 1930s, they caused a scientific uproar in England and the United States. Unfortunately, Wegener wasn't around to enjoy the accolades. Dedicated to the end, he had died in a Greenland blizzard in 1930 while trying to prove that the Atlantic is growing.

Supporting evidence for Wegener's theory trickled in. Researchers found remarkable similarities in rock sequences on both sides of the Atlantic. Then they discovered glacial deposits on both sides that had been overlain, at the same time, by coal. Those coal beds, thousands of miles apart, contain fossils of *glossopteris*, a tree-like plant with unusual tongue-shaped leaves.

Above the *glossopteris* fossils, on both continents, are desert deposits including sand dunes long since hardened into rock. And finally, at the top of the sequence on both continents lie dark rocks formed by cooling lava. It would have been a remarkable coincidence, say

---

1. Pangaea, which means "all lands," consisted of two huge landmasses; Laurasia to the north (Eurasia and North America), and Gondwanaland (land of the Gonds) to the south. The Gondwanaland supercontinent included South America, Africa, peninsular India, Australia, and Antarctica. The two continents were divided by a sea, roughly where the Mediterranean is today, which we call the Tethys Sea.

geologists, for precisely the same sequence of rock-forming events to have taken place on such widely separated continents.

How did *glossopteris* spread from one continent to the other? There is no way that its heavy seeds, the size of a pea, could have blown thousands of miles across the sea. *Glossopteris* is proof that those two pieces of land were once joined together. It's like putting the pages of a torn newspaper together—the lines read across.

New discoveries in paleomagnetism helped prove the theory. Instead of pointing north or south, magnetism in ancient rocks often points in varying directions, which means the earth's crust must have moved. Even so, Wegener's theory languished on the shelf of disbelief for close to 50 years.

Wegener believed the continents could slide around freely because the seafloor is flat. After all, dirt and debris have been pouring into the seas for billions of years.

The seafloor *must* be flat.

But in the 1940s, as scientists invented new ways to peer beneath the seas in response to wartime needs, they made a shocking discovery. The seafloor is not even close to being flat.

Huge underwater mountains with giant cracks in their tops rise from the very depths. Popping out of the seafloor one after the other, they create a continuous world-encircling mountain range of incredible proportions. Drain away the water and you'd see the largest mountain chain on earth, bigger than the Alps and Himalayas put together.

The Atlantic, Pacific, Arctic, and Indian Oceans are divided by huge underwater mountain belts running north to south for thousands of miles. Called rises (they *rise* from the seafloor), the submarine mountains tower over the adjacent watery dreamscape.

Shorter mountains rise 2,500 feet above the seafloor. Taller ones climb an unbelievable two and a half miles. Looming "only" one and a half miles above the surrounding countryside, Pikes Peak would look puny beside one of these submarine monsters.

Almost 800 miles wide at the crest (Cape Cod to Milwaukee), and as wide as a continent at the bottom (several thousand miles), each mountain chain stretches more than 6,000 miles across the bottom of the sea. Put together, they comprise the largest raised topographic feature on earth and account for approximately 23% of the earth's total surface area.

For sheer size, the Mid-Atlantic Ridge is clearly the winner. It snakes south through the centers of both the North and South Atlantic Oceans, splitting them almost exactly in two. Circling the southern tip of Africa, it then turns north into the Indian Ocean, twisting around the globe like a raised seam on a baseball for more than 15,000 miles. If it could be straightened it would extend more than halfway around the world.

It's the strangest mountain range on our planet. Strange, because every rise has a valley, called a median valley, or rift valley, running through its middle. Deeper than the Grand Canyon (more than a mile deep), and 8 to 30 miles wide, the valley runs up the side of every mountain, splits each summit in half, then continues down the other side for the entire length of the system. (The Mid-Atlantic Ridge has a rift valley; the East Pacific Rise does not.)

The submarine ridges and their cracks extend thousands of miles beneath the sea, then continue into and through the continents. The Red Sea Rift, for example, runs north into the Jordan Valley, while the East Pacific Rise zips into North America between mainland Mexico and Baja California, then continues northward into the western United States. The cracks seem to prove that the seafloor is pulling apart.

The surprises kept coming:

- Seismologists noticed that almost all earthquakes occurred at the mountain crests—but they didn't know why.
- Geologists noted that sediments were extremely thin or even missing on the summits, but were thickened on the mountain flanks—but they didn't know why.
- Geophysicists detected abnormally high heat coming from the valley centers—but they didn't know why.

- And when geophysicists from Scripps Institution of Oceanography surveyed magnetism on the seafloor (by towing a magnetometer behind a ship), they found unusually strong magnetic fields above the mountain crests—but they didn't know why.

They found other magnetic anomalies, too. Polarity on the seafloor ran in stripes. Though magnetization was normal at the center of the ridges, it was reversed further away; then normal again. One band of seafloor pointed north, the next band pointed south.

Parallel to the mountains and thousands of miles long, the stripes range from about one kilometer to about 100 kilometers (60 miles) wide. On average, positive and negative stripes added together at their widest points total 36 kilometers (21 miles). Similar alternating stripes blanket every seafloor in the world.

To top it all off, rocks on the eastern side of the ridge were the same age as—and remarkably similar to—rocks an equal distance away on the western side. Again, no one knew why.

In 1962, geologist Dr. Harry H. Hess of Princeton University, the man who first noticed that mountain-building events, earthquakes, and volcanoes are related; the man who named the rough circle of active volcanoes and earthquake epicenters that girdle the Pacific Ocean "The Ring of Fire"; the man who had led the way in marine research for years, told us why.

The submarine mountains were underwater volcanoes.

New seafloor is created at the mid-ocean ridges, said Hess. Accompanied by the slow rumble of almost continuous earthquakes, new oceanic crust constantly wells from long cracks, or rifts, in the middle of the valleys to form a thin layer of basaltic magma on the ocean floor. The entire seafloor is made of basalt.

Hess's ideas were later revised and modified by two British geophysicists, Fred Vine and D. H. Matthews, to form the basis of most present-day seafloor-spreading assumptions.

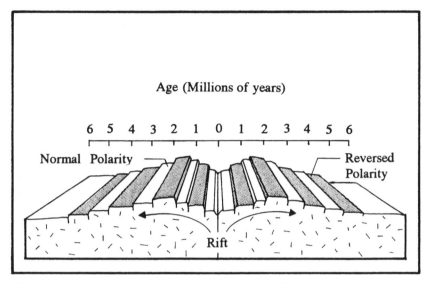

Newly formed basalt spews from long cracks (rifts) in the ocean floor, then crawls away in a conveyor-belt action from both sides of the rift. As it cools, it becomes magnetized in the direction of the earth's magnetic field of the time, thus creating alternating magnetic stripes.

Diving straight into the earth for 60 miles, the fiery grooves, called rift volcanoes, or linear volcanoes, are dividing lines between plates, and stretch the entire length of the chain. As basalt pours from the cracks, it quickly cools below the Curie temperature when it hits the water to become magnetized in line with the earth's magnetic field of the time.

That explained the heat. It was coming up through the cracks. It also explained the stripes. Positive stripes form when our magnetic field is normal, negative stripes form when it is reversed, leaving a permanent record on the ocean floor as to which way was north at the time they were created.

The stripes move as if on a giant planetary conveyor belt. First, molten magma pushes up through the crust. Then it inches upward and outward for about 1,500 meters (almost a mile), piling ever higher to form the ridges. Then it glides downward—about as fast as a fingernail grows—from both sides of the ridge toward the deep-ocean floor as if on separate moving sidewalks; sidewalks moving at identical speeds because the action is the same on both sides of the rise. The slope of its downward movement is so gentle that it appears to be flat.

For an impressive glimpse of this process of crustal formation at work, go to Iceland, where the Mid-Atlantic Ridge pokes its massive shoulders out of the frigid waters of the North Atlantic.

There, in that cold remote corner of the world, if you go to the right place at the right time, you can watch crust as it forms. As swarms of earthquakes jolt the immediate area, ground parallel to the ridge slowly spreads, sinking a few millimeters per year.

When tension and strain reach the breaking point, the land tears apart. Long cracks, or fissures, 6 to 60 miles long and about a foot wide, gape open in the tortured ground, and slices of crust drop down along the breaks.

Boiling lava, dark heavy lava, rises angrily out of the growling earth to fill the crevices. Each crack is only a small percentage of the entire rift system, and is only active every 200 to 300 years. But there's activity somewhere almost all of the time.

And that is how Iceland grows.

Icelandic basalt has a different chemical make-up, a different percentage of silica, than basalt from underwater rifts, but, above water or below, the growing and spreading action generally works the same way.

The South Atlantic has spread this way at a constant rate for the last 65 million years; about three quarters of an inch per year on each side of the ridge. At that rate, it moves about 12 miles, each way, every million years. But if the seafloor is spreading, where does it go? Is there a giant sea monster that scarfs it up for breakfast?

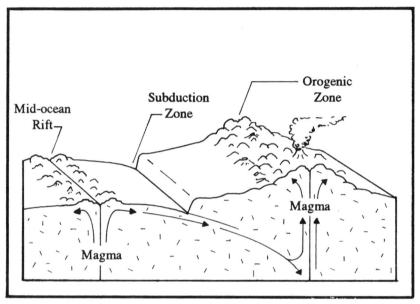

Rift, subduction, and orogenic zones.

Sort of. It's not a monster, of course, but the seafloor does get eaten. The same conveyor belt action that shoves it away from the submarine ridges carries it to deep trenches at the edges of the continents where it turns downward and dives beneath them. It's a conveyor belt straight to the furnaces of Hell. Then the seafloor is reheated, to be pumped from the ground thousands of years later by volcanoes.

But how could anyone be certain that the seafloor is indeed being eaten at the trenches? Joining forces aboard the *D/V Glomar Challenger,* a modified oil-well-drilling ship, scientists from several universities and institutions pooled their resources and talents to drill more than 200 cores into the deep ocean floor.

Called JOIDES (Joint Oceanographic Institutions Deep Earth Sampling), if they had found extremely old sediments, the newborn plate tectonic theory would have died in its infancy. Old sediments would have meant that the seafloor is neither spreading nor being eaten at the other end.

But the theory held up. Although our planet is more than four-billion-years old, no sediments more than 180 million years old appeared in the cores. And the closer to the trenches, the older the sediments.

Called subduction zones, or subsidence zones, trenches are about three miles wide and almost seven miles deep. Stack Pikes Peak on top of Mount McKinley, toss them both into a trench, and they'd still be half a mile short of reaching bottom.

As plates dive into a trench, they often snag and stop moving. Stuck together for hundreds of years at a time, but still shoving, incredible pressures build up. When they finally tear apart, they let go with a snap, and massive earthquakes rip across the planet.

A study of underwater volcanoes by J. Tuzo Wilson at the University of Toronto confirmed the JOIDES findings. Volcanoes get older, and show more erosion, the further they get from the rises. Crust *is* being eaten, in deep negative gravity areas (the trenches) near the edges of the continents.

At first, scientists assumed that the trenches plunged straight down. But in 1930, seismologists noticed that earthquakes occur in an odd pattern. Though most earthquakes (90% to 95%), and almost all deep earthquakes (called deep focus earthquakes), occur near subduction zones, some are deeper than others. The American seismologist Hugo Benioff decided to find out why.

Calculating earthquake hypocenters (the exact spot and depth where a quake begins) at different sites around the globe, he plotted their locations in three dimensions starting at the trenches and moving toward land. In 1954, Benioff published a landmark paper.

Earthquakes do *not* go straight down, he declared. Many earthquake zones plunge beneath the continents at an angle. The further from the trench, the deeper the earthquake. These angled underthrust zones are now called "Benioff zones."

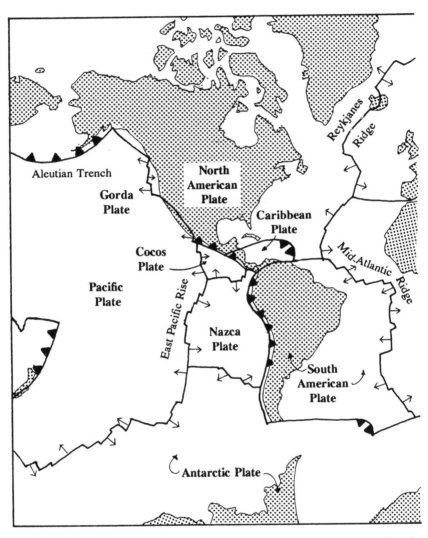

Simplified world map showing oceanic ridges, tectonic plates, and sub-duction zones. (After Forsyth and Uyeda, *Geophys. Journal R. astr. Soc.* Vol. 43, 1975.)

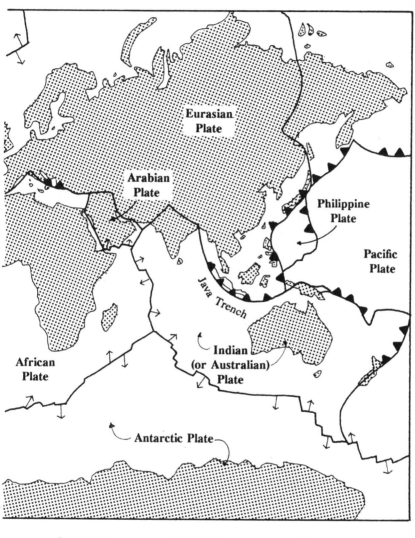

Eurasian Plate

Arabian Plate

Philippine Plate

Pacific Plate

Java Trench

Indian (or Australian) Plate

African Plate

Antarctic Plate

Continents   Direction of plate motion   Subduction zones

Japanese seismologist Kiyoo Wadati found a similar inclined pattern at the Japanese Trench off the east coast of Japan. As the seafloor wriggles beneath that island nation, earthquakes become progressively deeper, occurring up to 450 miles below ground and up to 600 miles west of the trench beneath Manchuria.

Tapping into the World Wide Standardized Seismograph Network (WWSSN) with its 110 seismograph stations around the world, scientists soon pinpointed the locations and angles of other Benioff-Wadati zones.

There are many. Plunging beneath the South American Plate along the west coast of South America, the Nazca Plate shoves the Andes ever higher into the sky. Another trench, the Middle Americas Trench, runs parallel to the coastlines of both southern Mexico and Central America. Meanwhile, the African plate is slowly being consumed by a system of trenches south of the Aegean arc where, scientists confidently predict, North Africa will someday collide with Greece and Turkey.

The Pacific and African plates are the size of continents. Others, such as the Juan de Fuca, the Gorda, the Caribbean, and the Persian, are much smaller. Pushed by the unknown forces that drive them, each plate rubs against, rams into, or dives under, other tectonic plates. The Gorda plate, for example, lurched about eight feet beneath the North American plate during an April 1992 earthquake in Northern California.

Slowly grinding northeastward across the floor of the Pacific under Mexico's coastline, the Cocos Plate slides beneath the western edge of the Caribbean Plate, causing almost all earthquakes and volcanic eruptions in Central America. Meanwhile, the Pacific Plate dives with glacier-like slowness beneath the state of Alaska.

Proving that the Pacific Plate dives at an angle beneath the 49th state, and not straight down, was left to Dr. George Plafker, a USGS geologist from Menlo Park, California.

On Good Friday, March 27, 1964, Alaska was shattered by the most powerful earthquake (8.4 magnitude) in its recorded history. A disaster for Alaskans, but the opportunity of a lifetime for geologists, the quakes gave them a laboratory where they could confirm their new tectonic theories; a chance to verify what their seismo-graphs had been saying all along.

It looked like the second coming of the Gold Rush, as a virtual army of scientists descended on the prostrate state. Among them was Plafker, an expert on south-central Alaska geology.

Plafker and his colleagues spent days flying over the wracked Alaskan countryside, looking down on a twisted, tortured land.

Then they got down on the snow-covered ground and walked. Step by relentless step they explored as many devastated areas as they could, looking for signs of uplift or subsidence and making detailed maps of surface displacement.

One of the oddest signs of uplift that they looked for was the bar-nacle line. Barnacles grow everywhere along the Alaskan coast. They grow on rocks. They grow on pilings. They grow on boats. They grow on oil rigs. They grow on almost anything that spends most of its time underwater. They grow to the mean high tide line—then they stop. They stop because they die if they're out of the water too long. The point where they stop growing, the barnacle line, is visible to the naked eye.

Though perhaps unsophisticated, the barnacle line became one of the quickest ways to detect uplift. If ground along the shoreline had risen, the barnacle line should have lifted with it. Old barnacles would have died and a new line of yearling barnacles would have formed below the old one.

And that's exactly what had happened. Plafker found places where the old barnacle line was 26 feet higher than before. The ground had risen two and a half stories!

He found something else, though, even more important. Since water is always level, barnacle lines should also be level. But some barnacle lines now sloped. Sloping barnacle lines proved that the ground had risen at an angle. It had to be dip-slip—underthrusting on an enormous scale. Immense planetary forces had driven a massive chunk of seafloor at an angle beneath the state of Alaska.

Benioff was right!

Benioff zones, we now know, are interconnected boundaries divid-ing the earth's crust into about 20 separate moving plates. The crust, or lithosphere (from *lithos,* the Greek word for *rock),* is 60 to 70 miles thick. In constant motion, the plates float on top of the asthenosphere (from *aesthenos,* the Greek word for *weak).*

The asthenosphere is 60 to 215 miles below the surface. As the plates move closer to the asthenosphere (about 50 miles down) they get hotter, turn plastic like hot tar, and ooze past one another in a fairly steady flow. But the cooler upper crust, the lithosphere, is solid and rigid. Unable to slide quite so easily, the plates snag. Then stresses build up. When the lock breaks, the sudden fracture causes earthquakes. That's why earthquakes occur almost exclusively at the edges.

The edges of the plates (called fault lines), interconnect with trenches and submarine ridges to continue all the way around the globe. The Juan de Fuca Ridge, for example, is a continuation of the East Pacific Rise offset by the San Andreas fault. It eventually connects with dip-slip (underthrust) zones in southern Alaska and the eastern Aleutians. Then it becomes strike-slip (the plates slide past each other sideways) on its never-ending journey to the west.

Unlike Benioff zones, strike-slip lines (or transform faults) *do* go straight down. Lying at right angles to the ridges, they're lines of pure slip where the plates glide past each other sideways like gigantic trains on separate tracks.

Since earthquakes occur mainly at the edges (the plates themselves are considered aseismic), most land deformation also occurs at the edges. Deformation occurs at underwater ridges where new crust is formed, it occurs at subduction zones where crust is eaten, it occurs at compressive folds where plates ram together to form mountains, and finally, deformation occurs, but not nearly as much, at the edges of plates (strike-slip zones).

Slashing diagonally across California from Mexico to Cape Mendocino north of San Francisco, the San Andreas fault is a good example of a strike-slip fault at work. Most of the Golden State lies on the North American Plate. But a 700-mile sliver on the western side of the fault lies on the Pacific Plate. As the Pacific Plate rotates away from North America it takes that sliver with it. Scraping sideways, it slowly glides north-northwest away from the rest of the continental United States.

Called right-lateral movement (if you stand on one side and watch the other plate move, it appears to be moving right), it has torn Baja California away from Mexico's mainland to create the Gulf of California. In Hollister, California, the motion is all too obvious. Residents watch helplessly as their land slides sideways about half an inch per year, and their curbstones, foundations, even entire warehouses, crumble and tear apart.

During the past 23 million years, some experts say, the west side of the fault has slid 210 miles. Others insist it has traveled 180 miles in the past five million years alone. In 18 million more years, depending on whose figures you use, Los Angeles will trade its smog for fog, having slid northwest to the outskirts of San Francisco.

In another 55 million years, L.A. will be a lonely island hundreds of miles west of Seattle, and after another 45 million years will plunge beneath the Aleutians and sink slowly into the mantle. Drop your trash into a subduction zone and you won't see it again for 100 million years . . . the biggest dispose-all in the world.

During the Great San Francisco Earthquake of 1906, land north of the city in Marin County slid up to 21 feet. Even more dramatic was the Fort Tejon quake (near L.A.) of 1857, when the ground slipped sideways up to 33 feet. Streams now run downhill, zig 33 feet, then run downhill again.

Thirty-three feet from strike-slip movement. Thirty-three feet from the kind of movement that "doesn't cause much deformation." To see *real* deformation, watch a head-on collision between plates: look at the Himalayas, northwest of India in Pakistan, which were folded and compressed when India rammed into Asia some 38 million years ago. (The word "compressed" is misleading. You'll see why later.)

That ancient two-continent collision is still at it today. Fifteen hundred miles long and 125 to 250 miles wide, the Himalayan range contains some 30 peaks rising 24,000 feet into the sky. Mount Everest, at 29,028 feet, soars more than five and a half miles into the cold and rarefied atmosphere. Now *that* is deformation!

Moving northward at two inches per year, India still pushes the Himalayas ever higher, causing earthquakes in Tibet, China, and Pakistan.

And it's all caused by continental drift.

The continental drift (or plate tectonic) theory is now considered fact, and satellite measurements confirm it. Using a fixed network of lasers, the Laser Geodynamics Satellite (LAGEOS), launched in 1976, confirmed that tectonic plates do, in fact, move. (Christodoulidis, Torrence, and Dunn, 1985)

Even our oceans and mountains are transient, repeatedly growing bigger or smaller, pushed or pulled by forces we still don't understand. It's a galloping globe that we live on. Europe is 40° closer to the North Pole than during the Permian, said Professor S. Warren Carey, professor emeritus of Geology at the University of Tasmania. Greenland is 50° closer, Siberia is 17° closer, the coastal region of Asia is 23° closer, and Australia is 45° closer.

"Every kind of study that has been made," said Carey, "palaeoclimatic, palaeomagnetic, tectonic, agrees that Australia, South America, Africa, India, Europe, North America, Greenland, are all in more northerly latitudes than they were in the Upper Palaeozoic or Mesozoic." All continents except for Antarctica (which seems to be temporarily anchored to the bottom of the globe—no one knows why), are headed north.

They're also headed east and west.

In 1951 when Japanese geologist K. Ichikawa found pebbles in the Kitikami Mountains that came from where the deep Pacific now lies, he realized that they came from Kuroshio and Oyashio, lands that once existed on the oceanic side of Japan. (Westerners call those lands North America.)

Same on the Pacific Coast of South America. During the Devonian more than a million cubic kilometers of micaceous sand was deposited on Bolivia, Peru, and northern Argentina. The sand came from a continental source where now there is no continent, only the deep Pacific. Additionally, fossils on one side of the Pacific show strong affinities with Devonian fossils on the other side.

"Where is this lost continent?" asks Carey. "It is Antarctica and Australasia, then a single continent, now separated from South America by the opening of the Pacific Ocean."

Same in northeastern North America. "Tens of thousands of cubic miles of quartz-rich sediments derived from the east were deposited on the east side of the Appalachians in the Late Devonian," said the Canadian geophysicist J. Tuzo Wilson, "A huge sediment source existed to the east of the mountains where now there is only water, then moved away."

Not only do plates move, said Wilson, they sometimes reverse direction. They slam together to form bigger landmasses, then pull apart. Constantly on the move, the world's continents continually split apart and move east, west, or north by thousands of miles, only to crash into each other again in slow motion at the other end, welding themselves back together as new and different Pangaeas. Ocean basins were thus born, grown, diminished, and closed again only to open one more time. To honor Wilson's pioneering work, that open-and-close cycle in the seas is called the Wilson Cycle.

Strangely, there appears to be a direct link between seafloor spreading, extinctions, and magnetic reversals. Some changes in plate movement direction, such as in the early Ordovician and Late Jurassic, took place at extinctions. And some occurred at times when magnetic reversals had speeded up.

So. We know where the plates come from, and we know where they go. We know how fast they move, and we suspect that magnetic reversals may somehow affect the movement.

But how?

7

. . . .

THE TRAITOR WITHIN

. . . .

Go north, young man, said the Mama proton to her kids. And north they went, billions of them, just like the protons in cosmic rays. Remember those cosmic rays we talked about earlier? cosmic rays filled 81% with protons? protons that spiral around the earth's magnetic lines of force before precipitating toward the North and South Poles? If the protons in cosmic rays move toward the poles, shouldn't earthbound protons do the same thing? I think they do.

Not only in cosmic rays, every speck of matter on this planet is made of protons. No more football-like spirals through the sky at thousands of miles an hour for these babies. They've been grounded. The game is down and dirty now, a game played out in the mud.

And a hard game it is, sloshing ahead at three quarters of an inch per year. Like an invading army marching with lock-step precision, protons (sub-atomic particles, really) move slowly, never-endingly, across the globe, pushing, shoving, bullying, bulldozing everything in their paths.

Look at what the earth is made of.

It's made of atoms. And atoms are made of protons and neutrons with minute electrons orbiting a central nucleus. Protons have a positive electric charge, electrons have a negative charge, and neutrons, though they behave like magnets, have no charge at all.

And that's the point. Protons, electrons, and neutrons behave like magnets; miniature magnets constantly trying to align with the earth's magnetic field, constantly moved by the same power that moves the needle on a compass.

Indeed, don't compasses prove all by themselves that magnetic forces can, and do, make things move? In a word, yes. Magnetic fields, said Robert Noyes in *The Sun, Our Star*, do very slightly perturb the energy levels within atoms.

And that's the secret.

Electromagnetic forces are the real movers and shakers of our planet. Electromagnetic forces cause all tectonic movement. Let's face it, *something* drives the continents across the globe. That something is billions of tiny magnets, all crawling north like miniature tadpoles in the mud.

And as they move, they take the continents with them. "When plate tectonics was first proposed three decades ago," said Stanley Chernicoff in his 1995 book *Geology*, "the plates were viewed as passive hitchhikers on the flowing asthenosphere. Recent findings, however, suggest that plates may contribute actively to their own mobility." Electromagnetic forces provide that mobility.

It works in our laboratories, why not in real life? When an electric field is applied to certain solutions of molecules (a molecule consists of one or more atoms), said B. R. Ware, "each particle moves toward the electrode of opposite polarity."

Called electrophoresis, the process is used to separate human blood plasma molecules into separate classes (alpha, beta, and gamma globulin). It's also used in DNA studies. (*Encyclopedia of Science and Technology*, 1992)

Since small electric currents, called electrotelluric currents, constantly flow through the soil (more about electrotelluric, or telluric currents, later), the same process, I believe, must occur in the earth itself.

Electrophoresis, I submit, drives the continents across the globe. It drives them south during times of predominantly reversed polarity and north during times of predominantly normal polarity.

But if all continents are headed north, what will happen when they get there? What will happen as more and more protons come nose-to-nose with their evil twins coming up from the other side of the globe? With billions more protons constantly pushing from the rear, where can they go? They can't stop, they can't go forward, and they can't back up. What to do?

They take a dive. One chunk of land dives beneath the other (we call it subduction), or it moves to one side (strike-slip), or climbs into the sky. Why? When those miniature magnets in the soil meet face-to-face with their brethren coming up from the other side of the globe, they repel each other, as all like-poled magnets must.

Thus begins a chain reaction of revolving chunks of land. As each chunk dives, it embarks on a collision course with other chunks hidden deep in the earth. Twisting to avoid colliding, it runs into others and twists yet again. It's the closest thing you'll ever see to a perpetual motion machine.

The North Pacific plate rotates counterclockwise, as do the North American plate and the Siberian block. Even countries rotate. Spain, for example, has rotated about 35°, said Allan Cox and Robert Hart in their 1986 book *Plate Tectonics, How it Works*. Islands rotate too. The Greek island of Zackinthos has rotated clockwise 26° in the last 11 to 12 million years alone. (Carlo Laj *et al.*, 1988) *Something* caused that rotation. Electromagnetic forces.

India has rotated 70°, Corsica-Sardinia has rotated 90°, Seram 100°, Mesozoic Mexico 130°, and Italy 110°. *All* plates rotate, said

Professor S. Warren Carey in his *Expanding Earth Symposium*. The earth's crust is made of rotating polygons inside bigger polygons inside bigger polygons, all jostling independently during earthquakes. "Adjacent polygons," said Carey, "often have different orientations like magnetic domains."

They're not *like* magnetic domains, Professor Carey, they *are* magnetic domains, dormant magnetic domains, ready to awaken at the next magnetic reversal.

Which brings us back to those Scotty Dog magnets of our youth. Remember holding them by their southern ends, trying to force their northern ends together? Almost impossible, wasn't it? They both pointed up, or one pointed up while the other pointed down, or they both pointed sideways. They simply refused to touch. They repelled each other.

Tectonic plates do the same thing. They repel each other. Look at the ways in which tectonic plates move (next page), and see if you don't agree.

It all goes back to the sun. Remember that process on the sun called supergranulation, where bunches of rotating bubbles combine to form larger bubbles that rotate as units? They rotate as units because they're restricted from crossing into each other's territories by invisible magnetic barriers; invisible repellant electromagnetic forces.

The same thing happens on earth, I propose, but it's done with solids instead of gases. It's slow-motion granulation, caused by magnetic repulsion.

That's why you see slices of rock sticking out of the ground at strange angles. Look at almost any mountain, and you'll see ancient strata jutting skyward at a tilt . . . shoved there by slow-motion granulation.

Land can rotate vertically (like a ferris wheel), or it can rotate horizontally (like a merry-go-round). Even mountains rotate. In Arizona, for example, geologists have found mountains that once rotated clockwise (called dextral rotation), while a few miles up the road are different mountains that rotated counterclockwise (sinistral rotation). Scotty Dog mountains, I guess you could call them.

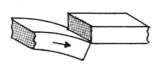

NORMAL FAULT
One plate has
dropped in relation
to the other

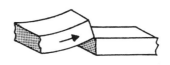

REVERSE FAULT
One plate has
risen in relation
to the other

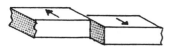

STRIKE SLIP FAULT
At least one plate
has moved sideways

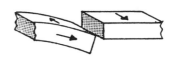

OBLIQUE FAULT
The plates have
moved horizontally
and vertically

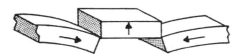

HORST
Ground between
faults appears to
have risen

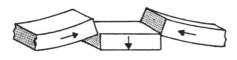

GRABEN
Ground between
faults appears to
have dropped

Repelling tectonic plates.

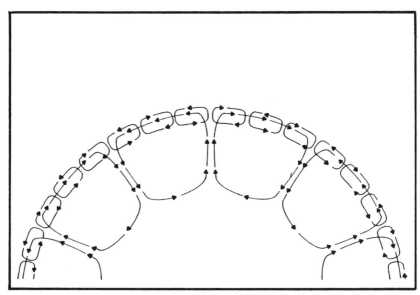

Revolving polygons.

Rotating chunks of land (polygons) can be as small as a few hundred yards across. Bunched together, they form bigger rotating polygons as small as three miles across. The three-milers combine into larger rotating polygons up to 40 miles across. And so it goes, ever upward, ever larger, until they grow up to be continents.

Each polygon is separated from its neighbor, says Professor Carey, by cracks that appear only during earthquakes. The cracks allow it to tilt independently and at varying angles. That's how earthquakes can destroy one side of a street but not the other.

The cracks [faults] prove that the earth is expanding, says Carey. But if faults are mere cracks in the ground, why don't they meld together as they get deeper? Both the San Andreas and Hayward faults go straight down for at least eight miles, maybe more. What keeps them from fusing? Repellant electromagnetic forces.

That's why we find magnetic anomalies at fault lines, anomalies so pronounced that they can be detected from the air. That's why we find sharp discontinuities in electrical conductivity at fault lines.

Seismic electric signals, said Haroun Tazief in his 1992 book *Earthquake Prediction*, sometimes become isolated by faults.

But what does this have to do with magnetic reversals? Don't chunks of our planet thrust, heave, subside, and rotate all of the time, not just at magnetic reversals?

Of course they do. But remember, tectonic activity goes berserk during magnetic reversals. Remember, two million cubic kilometers of basalt spewed out of the Deccan Traps at the K-T polarity flip. Large portions of the East Pacific Rise poured into the sea at the same reversal (see page 128). *Something* caused those eruptions.

Magnetic reversals, the traitor within.

Magnetic reversals shock the earth, triggering volcanoes and earthquakes around the world. Tectonic plates lumber across the globe to climb into the sky as newborn mountains, or dive into the ground in mammoth episodes of subduction, while huge chunks of land called thrusts charge across the face of the earth.

What are thrusts? They're giant rock formations that once moved sideways across almost level ground. Small thrusts exist in the Appalachians that have moved only three to twelve feet, while the Grand Banks (a series of underwater plateaus stretching from southeast of Cape Cod to southeast of Newfoundland) contain larger thrusts up to half-a-mile wide that have traveled for miles.

But the thrusts in western North America dwarf their eastern counterparts. Southeastern Idaho and western Wyoming boast huge thrusts that have bulldozed over existing land, even over the tops of previous thrusts, for 20 to 30 miles. (Picture a slice of land 20 miles wide, one mile tall, and as long as the state of Vermont. Now see it screeching sideways across New Hampshire.)

The Canadian and northwestern Montana Rockies contain a succession of such thrusts, each of them 10 to 20 miles wide and 50 to 150 miles long.

New Mexico and Arizona contain many thrusts, as do Colorado, Nevada, Texas, and California. In Utah, the Charleston-Strawberry-Nebo Thrust in the Wasatch Mountains, almost five miles thick, has traveled 25 to 120 miles eastward. Five miles thick? That's the size of a mountain!

Amazingly, some thrusts even moved uphill. Montana's Chief Mountain, for example, "traveled across the plains and climbed the slope of another mountain and settled on top of it." (Immanuel Velikovsky, *Earth in Upheaval*, 1955)

What drives the thrusts?

They're driven by gravity, many geologists believe.

Sure. Are we to believe that a chunk of land the size of a mountain and the length of a state crawled sideways across almost level ground for 120 miles—due to gravity? Impossible. Try using gravity to push a flat rock across a level field. Try using gravity to push your car sideways when you run out of gas. Gravity fights sideways movement. It's called friction.

No, gravity won't push anything sideways. *But magnets will.* And that's the secret. Those thrusts were driven sideways, I submit, by electromagnetic forces deep in the earth. We don't see it, because most of the movement occurs during aborted or full reversals.

Think back to the things that happen on our planet during solar flares and sunspots (magnetic reversals on the sun); things such as the northern lights, geomagnetic storms, and fluctuating magnetic fields. Now recall what those fluctuating magnetic fields do.

They create large electric currents on earth.

If a magnetic reversal 93 million miles away can create large electric currents on earth, doesn't it seem that a reversal just zero miles away could create *huge* electric currents?

We've been aware that magnetic reversals can produce electricity for more than 150 years, ever since the self-taught English physicist Michael Faraday built the first successful dynamo, thereby showing that a fluctuating magnetic field can produce an electric current.[1] It's Faraday's Law, and it's a basic law of electromagnetics: A fluctuating magnetic field can produce an electric current.

---

1. Faraday not only discovered electromagnetic induction, he established the foundation of almost all electromagnetic theory. He formulated the laws of electrolysis (Faraday's Laws), and coined many electrical terms still in use today, such as anode, anion, cathode, cation, and electrode. The unit of capacity, the farad, is even named after him.

If a man-made dynamo can produce an electric current, doesn't it seem that a fluctuating magnetic field as big as the earth itself might produce a planetary-sized current, sending billions of volts of electricity surging through the soil?

Now we know why those brontosaurs' tails were twisted and contorted over their backs as if they died in agony. They did die in agony . . . by electrocution.

What else could have possibly baked that mud so fast as to turn it to rock in an instant? What else could have possibly baked that mud so fast as to preserve footprints from 65 million years ago? What else could have possibly baked that mud so fast as to preserve ripples at the bottoms of ancient lakes, and the imprints of thousands of prehistoric raindrops?

That's right, raindrops! Indentations of thousands of tiny raindrops have been found in end-Cretaceous rocks from the Grand Canyon to Rocky Hill, Connecticut.

Imagine the kind of raw power that had to have been racing through our planet to have baked the mud so fast. That's the kind of raw power that could blow the top off a mountain.

And that's exactly what it did. Every volcano on the face of the earth jumped into frantic violent action, spewing billions of tons of ash into the skies and millions of kilometers of lava and basalt onto the ground and into the seas. (Scientists agree. Volcanic activity, both underwater and above, increased dramatically at the K-T.)

Underground, hot molten magma intruded into every crack and crevasse it could find, forming new igneous rock. (Again, scientists agree; the dates of many igneous intrusions are centered around the K-T boundary.)

As the volcanoes grew, so did the mountains, and the plains gave birth to the Rockies. Rugged young mountains elbowed their way up through the tortured soil from Mexico to British Columbia in a grand series of mountain-building events known as the Laramide Orogeny; a savage onslaught of a heaving earth, caused by magnetic reversals.

Raindrops and dinosaur footprints at Rocky Hill, Connecticut. The animal walked from a firm mud bank where rain prints are visible, into the soft ripple-marked bottom-mud of a pond or stream. (Photograph courtesy of the Peabody Museum of Natural History, Yale University.)

That's what jump-started the plates, I submit, and that's how many of the world's metamorphic (heat-altered) rocks were formed. Powerful surges of electricity raced through the sediments, transforming them into metamorphic rock in an instant, the same way that heat turns clay into brick.

Previously recrystallized sediments, re-molten and now re-cooled, recrystallized *in situ* (on site) one more time, changing composition yet again. (If they've been heated more than once, they're called polymetamorphic.)

Clay sediments will metamorphose (change) into slate at 100° to 200°C. At higher temperatures they'll metamorphose into schist and gneiss. And if they get hot enough they'll flow. Schist and gneiss have often flowed for many tens of miles. A funny thing about metamorphic rocks, say geologists, is that rotation of their mineral grains is common. Blame a reversing magnetic field.

Different kinds of metamorphic rock can form from the same clays, depending on what minerals they contain. With enough heat, pure limestone will recrystallize into marble, while clean sandstone will turn to quartz. Phyllite is a metamorphic rock. So are amphibolite, blue-schist, and granulite. Even granite, some experts insist, is made of heated recrystallized sediments. *Something* caused that heat. That something was electromagnetic forces.

It shouldn't take all that much. Utility workers tell of seeing downed power lines dance around with so much electricity spewing from their ends that they turned solid concrete into glass. It's not such a great stretch of the imagination to believe that a world filled with raging electricity could turn a few layers of clay into slate.

Nor is it a great stretch to believe that a world filled with raging electromagnetic forces could trigger volcanic eruptions. Indeed, magnetic and electric fields near volcanoes, said Robert and Barbara Decker in their 1981 book *Volcanoes*, often oscillate prior to eruptions. That's why volcanic eruptions are often accompanied by lightning; they're escaping electromagnetic forces.

Look at Iceland. During the birth of the volcanic island of Surtsey observers saw enormous purplish lightning bolts ripping through the eruption cloud, while thunder rolled and cracked above their heads.

Or look at Mount St. Helens. Eyewitnesses saw huge lightning bolts probing *upwards* thousands of meters during its 1980 eruption.

Or consider Westdahl volcano on Unimak Island in the Aleutians. Dormant since the 1970s, when it surged to life in November 1991 it treated residents of False Pass, a small town 60 miles northeast of the volcano, to a dramatic lightning show.

What causes volcanoes? As the seafloor plunges beneath the continents, it gets heated to thousands of degrees, becoming hotter and lighter than the magma it's diving into. Turning into miles-wide blobs of molten rock shaped like upside-down teardrops, it slowly rises to the surface, say scientists, due to heat convection. Though the process may take thousands of years, the bubbles of magma (or plutons), stay hot enough to melt any rocks that get in their way. When they finally reach the surface they explode out of the earth.

Heat, in other words, causes volcanoes.

But I disagree. Heat does *not* cause volcanoes. Volcanoes would happen even if they were made of solid ice. Look at this headline:

"Photos Show Ice Volcanoes on Moon of Neptune," said the story. (*Seattle Times,* 14 Feb 1992) "Voyager 2 spacecraft's pictures reveal what look like three gigantic, ice-gushing volcanoes on Neptune's frozen moon Triton." The photos show geyser-like black volcanic plumes of nitrogen ice, nitrogen gas and carbon dust soaring five miles into Triton's sky, along with long cracks filled with oozing ice.

No, heat doesn't cause volcanoes. It may speed up the process, sure, but heat doesn't cause volcanoes any more than heat drives tectonic plates across the globe. The real question is, What causes the heat in the first place? Electromagnetic forces.

Electromagnetic forces could also explain how mountains grow. At convergent zones (where plates collide), mountains supposedly grow by crumpling. The crumple theory began in 1815 when Sir James Hall pushed two piles of paper together and watched them crumple into folds. Mountains grow the same way, said Hall, by compression.

I think Hall was wrong.

Lightning over Surtsey. Ninety-second time exposure. (Photo by
Sigurgeir Jónasson, 1 Dec 1963)

So does Professor Carey. "Alpine foreshortening is a myth," said
Carey. "Plate tectonic theory insists that Africa rammed against
Europe and crumpled up the Alps. It insists that India's drive to the
north crumpled up the Himalayas, and that South America's northward drift crumpled up the Greater Antilles."

But that's not true, said Carey. The Alps have widened, not short-
ened. The Andes are widening too, and have been at least since the
Miocene. Mountains rise steeply out of the preorogenic trough, then
turn over and flow outward as thrusts, usually overriding previous
thrusts, sometimes slamming into their backs. Some thrusts have traveled up to 300 miles in a single epoch.

At one time the thrusts ended in sheer klippen (German for cliffs) that have since eroded away. "A peculiar thing about these klippen," said Carey, "is that they are upside down. They are less strongly metamorphosed at the base than they are higher up."

Good! That's how they should be.

Try it yourself. Lay two pieces of typing paper flat on a table. Now shove them together. Do they crumple? No. Nine times out of ten, using each other for support, they'll rise into the air with no crumpling whatsoever until they flop over—like thrusts—and head outward again.

Mountains grow the same way. They climb into the sky like those pieces of typing paper. Continents do *not* crumple into mountains when plates collide. It's their *refusal* to collide that's important.

The action begins deep in the earth where electromagnetically-driven plates collide. Continually shoving, but at the same time repelling, the plates get so hot at their edges that the earth begins to flow, pushing everything above it toward the sky. Then it moves outward. That's how thrusts are formed. Thrusts are the earthly equivalent of solar flares and prominences, flung upward and outward during magnetic reversals.

That's why klippen are more metamorphosed at the top. As more and more land pushes from behind, the parts closest to the collision point get the hottest. When the thrusts turn upward and outward they're upside-down, with the hottest areas on top.

The typing-paper analogy could also explain another mountain mystery. With the weight of the entire mountain above it, the base of every mountain "should" be highly compressed and unbelievably dense. It should be so dense, say geologists, and its mass above the horizon so attractive, that a plumb bob set up near a mountain should be pulled slightly toward it.

But when Sir George Everest, Surveyor General of India, began his first triangulations of the country, he was astonished to find that the Himalayas weren't pulling their weight. His plumb bob deflected toward the mountains by only a third of what it should have. The base of the Himalayas was less dense than the mass above it. The Pyrenees also have low-density roots. Indeed, mountains generally are supported by less dense material below them.[1] But if the sides of mountains lean against each other for support like those two pieces of typing paper, there's no need for them to be dense at the base. They're holding each other up.

Submarine mountains work the same way. Orogenic zones and underwater ridges, says Carey, are genetically identical. Both have zones of high heat flow restricted to two narrow linear bands, and both have land rising from their centers which then crawls away in a conveyor belt action. When the land reaches the end of the belt, it sinks back into the ground, rises upward, moves sideways, or simply falls off the end.

For a glimpse of the conveyor belt at work, listen to Robert Atwood describe Alaska's Good Friday earthquake. (Atwood, publisher of the Anchorage *Daily Times*, was in his modern home perched on a high bluff overlooking Knik Arm. He ran outside the minute he felt the first strong tremors.)

"The world went crazy," said Atwood. "The earth just opened up and I was going down. It seemed an awful long distance. But when I lit I was in soft, dry sand and I was at the bottom of a rather sharp V which opened up—kept opening up more. And trees, stumps, fence posts and frozen soil like boulders came rolling down on top of me. I had to scramble to stay on top to keep from being buried.

"I clambered up the side of the chasm that I was in and looked around and in every direction I just saw nothing but desolation. Houses at all angles and topsy-turvy and everything silent except for the last snapping and breaking. My house was nothing but kindling. It was out on the beach; it had been about ninety or a hundred feet above the water and it was now at sea level." (*On Shaky Ground*, by John J. Nance.)

---

1. First discovered under the Andes by Pierre Bouguer, such density quirks, called Bouguer anomalies, remain unexplained to this day.

What mysterious force of nature could make the ground slide downhill ten stories and sideways a fifth of a mile?[1] What mysterious force of nature could make rocks, trees, houses, even entire sections of cities, roll sideways across almost level ground into V-shaped holes that hadn't existed just moments before?

It's called liquefaction. Working deep underground, liquefaction grabs layers of sand and soil which are normally solid and shakes them until they turn to pudding. Then the entire mass begins flowing like mud toward the nearest low spot.

In Alaska, entire sections of land flowed into nearby streams from both banks, compressing bridges and buckling them in the middle. Same in Japan. A hundred years ago, liquefaction pushed the sides of several Japanese valleys together during the greatest inland earthquake ever to strike that island nation.

Now, anything that can shove both sides of a valley together is strong stuff, I'll admit. But is it always due to liquefaction? No. It may look like liquefaction. It may even *be* liquefaction toward the end of the quake. But that's not what caused the problem to begin with. When Atwood's house rolled down the hill, or rather, when his entire hill rolled down the hill, it was the conveyor belt at work. And it will happen again and again in the same places.

Leave the north country now, and jet south to the historic British colony of Port Royal, Jamaica. Jutting into the water on a steeply sloping peninsula, Port Royal was rocked to the very depths of its foundation during an unexpected earthquake on June 7, 1692.

Within minutes upper layers of sand and soil tore loose and slid toward the ocean taking Port Royal for a wild unwelcome ride. Like a modern-day Atlantis the town sank into the sea, in some places smothered under five stories of water.

There it sat until 1959, when American archaeologists investigated the disaster. Probing under six to ten feet of silt they found an

---

1. Atwood's house had crawled 1,280 feet. A large portion of his subdivision moved, too. A 130-acre chunk of flat land crawled 700 feet across a 2.5 percent slope and into the sea. So much for building on flat land. (Tom Alexander, *Smithsonian*, 1975)

astonishing pattern of destruction. Though many buildings were demolished, entire blocks remained virtually intact. Buildings stood with their walls erect and with their doorways and windows right where they should have been. In one dining room were stacked pewter plates, glassware and crockery.

How could an earthquake powerful enough to carry entire city blocks into the drink be gentle enough to leave glassware stacked on the shelves? Liquefaction? Or was Port Royal carried into the sea by a conveyor belt? A conveyor belt, I say, an electromagnetic conveyor belt.

The edges of our cities fall into the sea, say scientists, because they were built on "poor soil." They *always* blame poor soil. Construct your buildings on compacted, engineered fill, they swear, and the problem will go away.

But poor soil is only part of the problem. The problem lies in nature, and nothing is going to fix it. If the San Andreas fault can jump 33 feet at a clip, if Seward can jump 47 (as it did during the Good Friday earthquake), what do we honestly believe will happen to cities on the edge of that movement? Nothing will stop the conveyor belt from turning. Eventually, those cities will roll over the edge, even if they have to go in one solid chunk.

Do we honestly believe that a force capable of shoving the Pacific Plate beneath the state of Alaska will be held back by a few measly yards of compacted soil?

Still not persuaded?

Let me leave you with this. "In Kodiak," said Bryce Walker in the 1980 book *Earthquake,* "about an hour before the earthquake struck, a magnetometer recorded several pronounced magnetic disturbances. Scientists could only speculate about the meaning."

What's to speculate? Electromagnetic forces drive the plates, electromagnetic forces cause earthquakes, and electromagnetic forces cause extinctions.

A shift in the magnetic poles could be accompanied by earth-quakes, floods and lava.

—BILL GATES

8

· · · · · · · · · ·

EARTHQUAKE LIGHTS
AND
CRAZY SNAKES

· · · · · · · · · ·

On December 16, 1811, one of the strongest earthquakes to ever strike the heartland of the United States tore through the frontier city of New Madrid, Missouri. Trees twisted and splintered as entire forests crashed to the ground, while huge cracks, "some so wide that no horse could jump them," yawned open in the frozen soil.

Shaking the sleep from their weary eyes (it was two o'clock in the morning), terrified townspeople leapt from their beds only to watch helplessly as their cabins and chimneys fell to the earth and the ground collapsed beneath their feet.

Collapsing, settling, slumping, whatever you call it, New Madrid sank one and a half stories into the ground, while entire islands disappeared beneath the surly brown waters of the Mississippi. Then the river rushed backwards as three-story-tall waves surged upstream, over-topped the river's banks, and gushed into newly sunken areas to create St. Francis and Reelfoot Lakes in Tennessee.

Happy Holidays, New Madrid.

It looked like the end of the world, or a battlefield, or Armageddon, said panicked settlers, as the air grew thick with sulphurous vapors and eerie flashes of light burst from the ground, illuminating the skies like distant lightning.

What in the world?

Earthquake lights! Earthquake lights in New Madrid!

Most people have never heard of earthquake lights, much less seen them. No one knows for sure what they are. They're not lightning. They're not northern lights. And they're definitely not figments of someone's fertile imagination. Earthquake lights are real.

Earthquake lights have been spotted all over the world, many times in California alone. They've been seen during a quake off Monterey Bay, they've been seen during a quake near Hollister, and during yet another in Santa Rosa.

They've also been seen in China. In 1974, when a swarm of tremors unexpectedly rocked the city of Haicheng in Liaoning, a highly populated province of Manchuria, Chinese authorities sent an army of specialists to the area. Running surveys, making seismograms, mapping fault lines, and installing geomagnetic monitors, they saw some amazing changes. Much of the area had risen, tilting as it rose. More breathtaking, though, was what they saw on their geomagnetic monitors. Haicheng's magnetic field strength was increasing!

As it increased, huge temblors tore through the city. Earthquake activity increased to five times more than normal during the first five months of the year. Just before Christmas, the pace cranked up even more. Worried officials at the seismological bureau announced that a quake of 5.5 to 6 magnitude would strike within six months.

One month later, earthquake activity increased. Wells bubbled, and rats and mice crept from their holes and wobbled around like drunks on a Saturday night. Misguided snakes awoke from their long winter's sleep, crawled into the open, and instantly froze to death on the cold icy streets.

Death. The ultimate price for a stupid mistake. But was freezing to death any worse than being ground into snakeburger by shifting slabs of frozen soil? Maybe the not-so-crazy snakes knew something that humans didn't, because on the morning of February 4, a drumroll from the gods raced across the land. More than 500 earthquakes rocked Haicheng in the next 72 hours, with a magnitude 4.8 quake as a flourish at the end of the drumroll.

Then all was quiet.

Not taken in by the lull, wary Chinese authorities knew the worst was yet to come. At two in the afternoon, a nearby military commander took bold and decisive action. "There probably will be a strong earthquake tonight," he announced. "We require all people to leave their homes."

Luckily, three million people listened. Though numbingly cold (4°F below zero), they snuffed out their fires and moved themselves and their animals into the open. Then they waited.

It didn't take long. Five and a half hours later, at 7:36 PM, the quake bore in like an avenging army. "Great sheets of strange light flashed across the sky in all directions, clearly witnessed by millions of frightened people." (*On Shaky Ground*, by John J. Nance)

The earth heaved. Tortured bridges twisted apart as roads crumpled and crumbled. Jets of water and sand (sandblows), shot 15 feet into the air. Nine of every ten buildings in the province were severely damaged or destroyed, and the cities of Yingkow and Haicheng almost completely demolished.

All in 120 seconds. Two minutes.

If that had been a "normal" earthquake in any other area of three million people, the number of deaths would have been astronomical. But in Liaoning Province, only 300 people were killed.

No one was naive enough to think earthquakes could be tamed, but it did look as if they could be predicted. With the Liaoning experience tucked under their belts, swaggering earthquake specialists turned their attention to Tangshan, a thriving industrial and coal-mining city of one million about 100 miles east of Beijing.

The region was ripe for the plucking. Though no recent earthquakes had occurred in Tangshan, suspicious scientists knew it was only a matter of time. A major fault ran directly beneath the city.

Smothering Tangshan with sensors and surveillance, their suspicions were soon confirmed. The familiar signs of impending disaster were everywhere. Gravity had changed. Electrical resistance had changed. And in January 1976 magnetism changed . . . sharply.

By the end of July, the earth looked as if it were run by committee. Indecisive groundwater levels moved up. Then down. Then up. Then down. Confused animals fared no better, turning into witless yo-yos unable to make a decision.

All the tell-tale signs were there. At 3:42 in the morning, July 28, 1976, a cold-blooded killer marched into Tangshan and stabbed it in the heart.

But this brazen killer didn't like working at night. One second it had been dark. Then it was bright, as if the sun had come up early that day, as if someone had flipped a switch in the heavens.

Daylight *snapped* through Tangshan. Mammoth sheets of white and red lights raced through the early morning sky, pulsating like an aurora borealis gone wild; lights that could be seen by startled citizens up to 200 miles away.

Earthquake lights; again, and again, and again!

Bright lights. Lights so bright that had they been flashing in London, they could have been seen from the banks of the Seine. Lights so bright that had they been flashing in New York, they could have been seen from Pennsylvania, New Jersey, West Virginia, Virginia, Washington, D.C., Delaware, Canada, and all six New England states. Lights so bright that they lit up more than 100,000 square miles of Chinese countryside. Lights so bright, said awed observers, that they looked like raging sheet lightning in a violent storm.

But there was no storm. No storm in the sky, that is. But a fero-
cious storm raged deep in the ground. Within seconds, the earth began shaking; shaking so savagely it seemed, that an angry vengeful
giant must be crouched beneath the city, pounding it with a mighty
sledgehammer. Measuring 8.2 on the Richter scale (*Newsweek*, 9
Aug 1976) it was the strongest earthquake anywhere since the 1964
Good Friday upheaval in Anchorage.

Hammering Tangshan with a spiteful fury, it threw thousands of
petrified Chinese high into the air, some so high that they bounced
off their ceilings. As they fell to the floor, their ceilings followed
them down, crushing them beneath a deadly rain of debris.

The city was under siege, and it lost the battle. Bombed from
within, fully 20 square miles of Tangshan were demolished. As its
buildings collapsed, so did the lives that had built them.

Trying to hide the extent of the disaster, faceless Chinese officials
put a lid on reporters' activities. But a boiling pot lifts its own lid.
Tragedies such as Tangshan refuse to be hidden. Up to 750,000 Chinese were killed, Western experts believe, and another 780,000 injured, in the deadliest earthquake in 400 years.

Do you suppose the bewildered dinosaurs saw any earthquake
lights before they died? The mystified giants couldn't have been
much more confused than we are. It's been 65 million years now,
and we still don't know what causes them. But though their cause
may be in doubt, there is no doubt that earthquake lights exist. It's
hard to deny what millions of Chinese saw with their own eyes.

It's also hard to deny photos. T. Kuribayashi, a Japanese dentist,
captured the elusive lights on film, taking 35 photos during an earthquake swarm in Matsushiro. His pictures show rocks, trees, and
mountains, all starkly silhouetted against a lighted Japanese sky.

Kuribayashi isn't the only Japanese to have seen the lights. The
1981 summer issue of *Earth Science* tells of two Japanese scientists
who collected earthquake-light reports from more than 1,500 eyewitnesses; reports of darkened skies suddenly lit up "as if by sheet lightning," or of lights which "resembled the northern lights, with
streamers diverging from a point on the horizon."

The lights lasted from ten seconds to two minutes (lightning normally lasts less than two seconds), and were usually spied not at earthquake epicenters, but at the tops of mountains (where the two sides lean together, I'll bet).[1]

Sometimes seen before temblors, sometimes after, earthquake lights are most intense at the height of a quake. And for some reason, they show up only during bigger earthquakes. They don't seem to affect magnetism, but do interfere with radio transmissions primarily in lower frequencies, and are most commonly seen during colder weather.

They also show up at geological fault zones in the unlikely guise of UFOs. A study of UFO sightings by French scientist Dr. Claude Poher found strong correlations between so-called UFOs and earthquake lights.

UFO sightings increase dramatically during magnetic storms, Poher found, especially near fault zones. Almost half of all sightings were made within 1,100 yards of a fault. One out of five sightings were made on top of the fault itself. Stoke-on-Trent, for example, a major fault zone in central England, generates an unexplainably high number of UFO sightings. In truth, said Poher, they're earthquake lights, escaping from fault lines during magnetic storms.

Others agree. Earthquake lights may be the source of some UFO sightings, said John Derr at the annual meeting of the Seismological Society of America in Santa Fe, New Mexico, in April 1992.

UFO sightings in New Mexico during 1951 and 1952 were clustered within 60 miles of three earthquakes that occurred less than a year later, said Derr, a geophysicist with the USGS.

After eliminating any sightings that could be written off as satellites, planets, airplanes, or meteorites, and eliminating any reports of contact with extraterrestrial beings, Derr still found 80 reported

---

1. Illumination of mountain peaks at night, called "Andes glow" or "Andes lights," is relatively common. The majority of such reports come from the Andes mountains of Chile, Bolivia, and Peru, but have also been reported in the Alps, Mexico, and Lapland. Electric fields above these mountains sometimes show a 25-times magnification. (Markson and Nelson, *Weather*)

sightings near the epicenters of impending quakes east of Tucumcari, northwest of Carlsbad, and near Los Alamos. (Interestingly, earthquake lights have been reported for years in Washington state's Cascade Range, in Missouri, and in other states near the New Madrid fault zone.)

"There is evidence that underground stress generates radio signals before some earthquakes," said seismologist Jim Mori, "so a similar phenomenon may explain earthquake lights."

Earthquake lights, many scientists now agree, are caused by electromagnetic disturbances strong enough to create large electric fields close to the earth's surface. But what could those disturbances be? Three researchers with the USGS in Menlo Park—James Byerlee, David Lockner and Malcolm Johnson—think they know the answer.

Rocks that go bump in the night.

As land movement occurs during an earthquake, they hypothesize, a thin strip of rocks rubs along the fault line and gets heated by friction. Thus heated, the rocks vaporize any nearby water. The air space forms an insulation barrier around the slipped section of fault, which creates an electric field, and *voila!*—earthquake lights. Earthquakes, in other words, cause electromagnetic disturbances.

But I think they've got it backwards. Saying that earthquakes cause electromagnetic disturbances is like saying that waterwheels make rivers flow. Electromagnetic forces *cause* earthquakes, not the other way around. The bigger the quake, the more likely that some forces will escape. That's why earthquake lights show up only at quakes measuring seven or more on the Richter scale.

Data from the 1989 earthquake in San Francisco lends support to this view. Prior to the quake, atmospheric scientist Antony Fraser Smith at Stanford University installed a device to measure radio signals in the upper atmosphere. Twelve days before the quake, radio energy levels increased to 30 times more than normal. Three hours before the quake, readings jumped to 300 times more than normal. This may be a way, said Smith, to detect earthquakes in advance. (Thomas Y. Canby, *National Geographic*, May 1990)

Even earlier, the American astronomer Warwick noted the correlation between radio signals and earthquakes during a magnitude 8.7

quake in Chile, when he detected a powerful electromagnetic emission from his observatory thousands of miles away in Hawaii. It might be a worthwhile gamble, said Warwick, to install radio receivers along the San Andreas fault. (*Science News*, 20 Mar 1982)

Another way of detecting earthquakes in advance comes from Greece. Called the VAN method, it was pioneered in the early 1980s by the Greek scientists Varotsos, Alexopoulos, and Nomikos (hence the initials VAN). They buried electrodes in the ground to track the low-voltage currents which continuously circulate through the soil, and to detect the seismic electrical signals that always, *always*, they say, precede earthquakes.

"Every sizable earthquake," says Varotsos, who heads the solid state physics center at the University of Athens, "is preceded by a seismic electrical signal (SES)." Inversely, "every SES is always followed by an earthquake the magnitude and the epicenter of which can be reliably predicted." (*Tectonophysics*, 1984)

They've had great success, pinpointing impending earthquakes to within 60 miles and to within six to 115 hours of when each quake occurred. In 1983, they successfully predicted 21 of 23 impending earthquakes. As they fine-tuned their methods, they became even more precise.

During the waning months of 1988 and into early 1989, they predicted 17 impending quakes. All 17 were confirmed in terms of both location and magnitude, said Haroum Tazief, in his 1992 book *Earthquake Prediction*. "I had been an absolute skeptic," said Tazief, a famous French volcanologist. "But by the end of 1988 I had been converted in a few months into the most enthusiastic proponent of the VAN method."

More recently, Varotsos claims to have successfully forecast the 6-magnitude quake near Thessaloniki on 4 May 1995, the 6.6-magnitude quake in northern Greece on 13 May 1995, and the 6.1-magnitude quake in southwestern Greece on 15 Jun 1995.

Some seismologists dispute the effectiveness of the VAN method, but Varotsos stands firm. "How many more people will have to die," he asks, "before this method is recognized as being correct?"

Another leading proponent of the VAN method is Seiya Uyeda, a professor at both the University of Tokai and Texas A&M University, who helped arrange a March 1995 tour of the Greek facilities by scientists from the Japan Meteorological Agency.

Three months later, Japan decided to invest heavily in the VAN method, budgeting more than ¥250 million (US$3 million) to set up 20 detecting stations around the Kobe area to measure electric currents in the earth. (David Swinbanks, *Nature*, 22 Jun 1995)

Though not well understood, naturally occurring electrical currents, called electrotelluric currents, flow on and through the earth's crust all the time. Generally flowing parallel to the crust, electrotelluric currents are frequently used by geophysicists to map subsurface structural features such as faults, sedimentary basins, and layered rocks. Anomalies in current density or gradient indicate a change in subsurface conditions. Electrotelluric currents are induced, says Theodore Madden, of the Massachusetts Institute of Technology, by variations in the earth's magnetic field.

If electrotelluric currents are induced by normal everyday variations in the earth's magnetic field, imagine what a full-fledged magnetic reversal would do!

Maybe we can ask the snakes.

You know, the crazy frozen snakes of Haicheng. Do they, and all of the other crazed animals, know something about electrotelluric currents that we don't? Behaving erratically immediately prior to earthquakes, their unpredictable actions have fascinated and frustrated scientists for years. Dogs, cats, cows, horses, and other barnyard animals, along with fowl, fish, even insects, often act as if they're going insane.

Stories about the demented behavior of animals prior to earthquakes have filtered down through the ages. But what scientist wants to admit that he listens to gossip? What scientist wants to repeat the "rantings of rural barefoot seismologists?"

How do you explain, scientifically, why birds almost too tired to fly refuse to land? Or why cockroaches twirl about in crazed indecision? Or why rats run from their houses and holes as if they're abandoning ship? Or why chickens and pigeons refuse to roost at sunset? Or why cats abandon villages, all just before earthquakes?

Even fishes go crazy. The sluggish catfish, normally a bottom dweller, sometimes becomes so agitated shortly before large earthquakes, say Japanese zoologists, that it actually leaps from the water. But why? Possibly, said Bryce Walker, in the 1980 book *Earthquake*, it's due to their extreme sensitivity to vibrations and electrical impulses. (According to early Japanese folklore, earthquakes were caused by gigantic catfish thrashing around in hidden underground streams.)

Animals and insects sense *something,* baffled scientists agree. "Many animal behaviorists," said Stanford University biologist Evelyn Shaw, "now believe that some animals are indeed sensitive to weak magnetic fields." Some songbirds, for instance, can sense changes in magnetism far below the amplitude of the earth's field.

Even magnets get into the act. Many years ago in Edo (now Tokyo), a huge horseshoe magnet hung in front of an optician's office. Doing what magnets are supposed to do, it attracted things made of iron. But two hours before a destructive quake in 1885, it temporarily lost its power and all nails and iron pieces dropped to the ground. (Tsuneji Rikitake, *Earth-Sci. Rev.*)

Earthquake lights, bubbling wells, crazy snakes, frenzied fish, cockroaches twirling in a circle, birds refusing to land, and magnets losing their power, all just before earthquakes.

If a handful of comparatively minor magnetic fluctuations can do all of those things, imagine what a full-fledged reversal could do.

Scientific thinking often benefits from the throwing of "bombs"—the publication of ideas so revolutionary that one half of the profession is scandalized, while the other half is captivated by the prospect of daring new solutions to old problems.

—ROBERT BAKKER

9

· · · · · · · · ··

DIVING TURTLES

· · · · · · · · · ·

The earth is built on the back of a giant turtle who floats on top of the sea. When the turtle decides it wants to get wet, it slips deeper into the water.

—ANCIENT INDIAN LEGEND

We laugh. But that ancient Indian legend is a lot closer to the truth than we like to admit. The sea doesn't rise, the land takes a dive. When it pops back up, as it always does, it only *looks* as if sea levels declined.

And that's what happened at the K-T. End-Cretaceous seas had risen so high that they covered a third more of the continents than they do today. The Rockies were beachfront property back then, with an ocean view to the east.

Then sea levels fell a fifth of a mile. But where did the water go? It didn't go anywhere. Not for long anyway. That old Indian legend had it right—the land rose.

Changes in sea level coincide with warping of the continents and orogeny (mountain creation) with "remarkable accuracy," said J. G. Johnson of Oregon State University. "The correspondence is so consistent in general, and even in detail, that it must reflect a fundamental relation." The Arbuckle and Wichita orogenies, the Ancestral Rocky Mountain building period, and the Palisades, said Johnson, all correlate with sea level fluctuations. (*Oros* is the Greek word for mountain; *geny* is from genesis, or creation.)

Sea level changes do occur in phase with rising and falling land, said L. L. Sloss and Robert Speed of Northwestern University. "Abrupt localized vertical movements of several kilometers have occurred simultaneously in the past all over the world."

That's how sea levels could fall in one place and not the other. Look at the mid-Jurassic, when sea levels fell in the North Sea while remaining the same in the rest of the world. "There must have been significant tectonic activity," said paleontologist Anthony Hallam, "both uplift and subsidence, in an extensive sector of Pangaea."

And what about those 180-foot-tall cliffs of water that we laughed about earlier (Chapter 3), when the seas retreated from Australia but remained unchanged around the rest of the globe? If Australia rose, the mystery is solved. "It doesn't necessarily indicate uplift of the whole country," said Hallam, "but movements on a regional scale are implied."

Or look at the K-T. Though the seas stood 1,100 feet deeper than today, they can account for only half of the sedimentary deposits found on the Gulf or Atlantic coastal plains or in the western interior. Some deposits measure almost two miles thick.

Where did the sediments come from? "The immense thickness of deposits," said Hallam, "indicates a high rate of subsidence next to the rising Laramide mountains."[1] As the mountains grew, in other words, the land beside them sank.

That's how it works. One section of land grows into a mountain while the one beside it sinks into a valley. And if it sinks too far, seawater rushes in to create an ocean. Entire islands sink beneath the sea in one part of the world while others rise like the Indians' mythical turtle.

How else to explain sunken mountains such as the Sweetwaters in central Wyoming? Once as tall as the Wind River uplift (also in central Wyoming), the Sweetwaters foundered. Then, except for their peaks, were buried by erosion from surrounding lands. Now they're slowly being exhumed in a new cycle of erosion.

Did the Sweetwaters sink, or was it that old diving turtle trick again? Did the surrounding land rise, making it *look* as if the mountains sank? I don't know. The point is, that various sections of land do dance up and down.

Land not only dances up and down, it does it in the same places. *And it rises and falls in a cycle!* "I see compelling evidence," said Professor S. Warren Carey, "that orogenesis is cyclic and pulsed." "Expansion waxes to a crescendo, then wanes perhaps to zero before the next wave of expansion." On average, major orogenic activity, including volcanism, occurred about every 30 million years.

How could this be? How could parts of a continent be rising while other parts are dropping? And why in cycles?

It's linked to our galactic orbit—and to magnetic reversals.

There is a broad but weak spectral peak in mountain growth, said Michael R. Rampino and Richard B. Stothers of NASA, which corresponds to the solar year. (*Science*, 21 Dec 1984) The late S. K. Runcorn at the California Institute of Technology concurred. Major peaks in both orogenesis and volcanism, said Runcorn, occurred about every 240 million years (plus or minus 60 million).

"It suggests a causal relationship," said Johann Steiner of the University of Alberta. "The driving engine must be extraterrestrial."

---

1. The Laramide mountains, which stretched from Mexico to Alaska, were precursors to today's Rocky Mountains.

Peaks in glaciation, Steiner added, also correlate with our galactic orbit. (*Geology*, Oct 1973)

Indeed, many scientists think that's where the water went, it got locked up on land as ice.

Take the end-Ordovician. Glaciation was so extensive at the end-Ordovician, says paleobiologist Steven M. Stanley, that ice reached into the Sahara Desert. As sea levels fell, the glaciers grew. When the ice melted, sea levels rose again.

Glaciation believers try to blame everything on ice, even rising land. Land around Lake Superior is rising 15 inches per century, said John and Katherine Imbrie in their book *Ice Ages: Solving the Mystery*. Land is rising on the other side of the globe, too, in Norway, Sweden, Finland, and Denmark. It's rising about three feet per century north of the Gulf of Bothnia, and about 16 inches per century near Stockholm. The greatest uplift, about three quarters of an inch per year, occurs in western Sweden.

And did you hear about the uprising in Texas? During Alaska's Good Friday earthquake, parts of the countryside near Dallas rose and fell almost four inches (*On Shaky Ground*, by John J. Nance). The South kept its word . . . it did rise again.

During the past 18,000 years, say Officer and Drake, parts of the United States' east coast have uplifted some 400 feet. Alaska's Middleton Island has risen 100 feet in the past 4,500 years alone (*Plate Tectonics and Geomagnetic Reversals*, Allan Cox).

Why is the land rising?

Because thousands of years ago, say ice believers, the weight of the ice pushed the land down. Now that the ice is gone, the land is slowly rising (updoming) to its original level.

Don't you believe it! Even if the land did sink earlier, it had almost nothing to do with the long-gone glaciers. The land rose because electromagnetic forces shoved it upward.

Glaciation theories simply do not "cut the ice." If land is rising because the ice is gone, why are parts of California rising? (L.A.'s Santa Susana Mountains, for example, uplifted about one foot during an earthquake in January 1994.) Why is Florida rising? Why are the Adirondacks rising? (Chernicoff, 1995) Why is the Chesapeake Bay area subsiding? (Officer and Drake, *Tectonics*, Dec 1985) Why is the Mississippi delta area subsiding at the rate of 15mm per year? (Dawson, 1992)

And why does land rise at an angle? Why is the northern end of Lake Superior rising, but not the southern end?

Because ice has almost nothing to do with it. "The movements are tectonic," said Sloss and Speed, "not responses to the weight of ice or mountains."

Same with changing sea levels. The dominant factor, said Anthony Hallam, was more likely to have been changes in the capacity of the oceans due to uplift and subsidence of oceanic ridges and subduction zones in conjunction with continental thickening.

Others agree.

Sea level changes may result from strong submarine igneous activity, said Alexander R. McBirney of the University of Oregon.

Sea level changes may be caused by increases and declines in underwater igneous activity, said James P. Kennett of the University of Rhode Island.

Sea level changes may be due to pulses of rapid seafloor spreading [underwater volcanism], said James D. Hays and Walter C. Pitman III of Lamont-Doherty Earth Observatory. A pulse of seafloor spreading during the Middle Mesozoic, they said, could have raised water levels as much as 1,500 feet. (Tom Alexander, *Smithsonian*, Feb 1975)

Aha! Now we're getting somewhere!

Underwater volcanism affects sea levels!

End-Ordovician seas, for example, rose during a time of massive seafloor spreading. And the Cambrian sea level rise, which created the Iapetus Ocean, also occurred during a major spreading episode, as did the sea level rise at the end-Triassic, which coincided with rifting all over the world (Hallam).

And it happened fast!

Some sea level rises occurred so fast, said Norman D. Newell, in *Megacycles: Long-Term Episodicity in Earth and Planetary History,* that "great freshwater swamps were flooded within a few hours or days, without a gradual transition in salinity."

All sea level rises may have been caused by rifting. "Changes in spreading rate on a global scale," said Robert E. Sheridan in 1986, "are probably a major cause of large scale (100-300m) eustatic sea level changes." *The rises appear to be related,* said Sheridan, *to the frequency of magnetic reversals.* (Emphasis mine.)

Same at the end-Cretaceous. Seafloor spreading was so vast at the K-T, said Fred Vine, that the East Pacific Rise, at least in the north and south, perhaps the length of the Pacific, was initiated at the time. (*Science,* 16 Dec 1966)

Perhaps the entire world!

Most of our present-day mid-ocean ridge system, said Maurice Ewing, one-time director of Lamont-Doherty Earth Observatory, was initiated at the end-Cretaceous.

It's a never-ending cycle.

As more and more basalt pours into the seas, the waters rise to flood the land. Then the seafloor dives beneath the continents shoving them ever upwards, at which point the continents shake the water off their backs and the process begins anew.

With ever more seafloor diving beneath them, entire continents crawl skyward. Five feet, 10 feet, 30, who knows how high they will climb? And as they climb, they form those three-mile-high cliffs at the edges of the continental shelves.

If there's enough power hidden in the heart of our planet to shove entire mountains into the sky, there should be enough power in there to easily shove a continent a few measly feet.

"The correlation cannot possibly be denied," said Dutch geologist J. H. F. Umbgrove. "There is a rhythmic character to orogenesis which is time related to transgression and regression of the seas, to formations of basins, to plutonic and volcanic activity *and to ice ages* [which] points to a common deep-seated cause."[1] (Emphasis mine.)

And it all has to do with magnetic reversals.

---

1. Clearly, ice ages and volcanism do go hand-in-hand. During the ten major European glacial stages of the past two million years, said J. R. Bray, "major volcanic eruptions apparently occurred at the crucial moments to have triggered each one." (*Science,* 15 Jul 1977)

Look at the isthmus of Panama, which rose from the sea about two million years ago at the beginning of the Pleistocene . . . at a magnetic reversal.

That's right! The isthmus of Panama rose from the sea at a magnetic reversal! The Cascades and the Coast Ranges in the northwestern United States rose at the same reversal.

Other episodes of rising land also occurred near reversals. Germany's Rhenish massif, for example, uplifted near the Brunhes reversal of 780,000 years ago. (H. Böhnel, 1987)[1]

Let there be no doubt in your mind: rising land, mountain growth, volcanism, and magnetic reversals are inextricably linked. Cores collected by the USNS *Eltanin* "show clearly," said oceanographers James D. Kennett and N. D. Watkins, that peaks in volcanism occurred *during* reversals. In 14 samples from eight different cores, only one did not show volcanism at a reversal. Mauritius, Rodriguez, and the Réunion Islands in the Indian Ocean all had volcanic activity at reversals. So did Nunivak Island.

So did New Zealand. Of five distinct layers of ash on New Zealand, said Kennett, one lies near the end-Jaramillo reversal, while two actually straddle the Brunhes/Matuyama boundary. Volcanic ash coinciding with the Jaramillo reversal has also been found in the North Pacific. "We find these observed changes," said Kennett, "difficult to accept as purely coincidental."

Almost all tectonic movement can be linked to magnetic reversals. Seafloor spreading, sea level changes, mountain growth, earthquakes, and volcanism, said Peter Vogt of the U.S. Naval Oceanographic Office, all seem to speed up whenever the frequency of reversals speeds up. The more frequent the reversals, the higher the discharge.

Lest you missed that, let me say it again: *The more frequent the reversals, the higher the volcanic discharge.*

---

1. A massif is a chunk of land, surrounded by faults, which rose or sank as a unit. The Rhenish massif is a huge plateau running from the Lahn River to the Belgium border. Bisected by the Rhine River, it's bounded by Bonn on the north and Bingen on the south.

Vogt's story begins with the Hawaii-Empire chain of islands, which were formed by volcanic hot spots deep in the mantle. As the seafloor drifted over a hot spot, lava spewed up to form an island. Millions of years later, as the seafloor crawled across the planet, lava burst up in a different place to form another island. That's why each island in the Hawaii-Empire chain is a different age. (The Hawaii-Empire chain is a string of islands and seamounts stretching some 3,600 miles across the Pacific, with the island of Hawaii being the youngest.)

Now, if the seafloor had always moved in the same direction, the islands should pop out of the water in a straight line. But they don't. Look at a map. They head due west for awhile, then take a distinct bend to the north.

Why the bend, Vogt wondered, and when did it form? It formed about 37 million years ago, he found, during a peak in magnetic reversal frequency. A sharp change in plate movement direction occurred at the same time at the Mendocino fracture zone.

Did plate motion change in the Atlantic too? An unqualified yes. A change in direction occurred at the Reykjanes Ridge south of Iceland. Different plates, different oceans, on opposite sides of the globe. And yet, both had switched directions during the same peak in reversal frequency. Both experienced volcanic activity, too. Vogt was on the trail of something big.

Like a bloodhound following the scent, he stuck his nose into everything. He studied the Keathley/Mercanton transition, he studied the Bermuda Discontinuity, and he studied changes in seafloor spreading south of Rockall Bank in the northeast Atlantic. All dates coincided. "Geomagnetic and tectonic events," said Vogt, "are recorded on the same tape." They also correlate with hairpin turns in the polar wander curve.

What is a polar wander curve? It starts with that point on top of the globe called magnetic north. Restless, always on the move, magnetic north constantly cruises and curves around the surface of our planet. The curves and arcs thus created are called the polar wander curve. But sometimes, instead of making long slow curves, it slams on the brakes, makes a U-turn, and heads back to where it began.

Those U-turns correlate with mountain-building episodes and plate movement reversals for almost two billion years. They can be linked to the Kibarian-Elsonian orogeny of 1.3-1.2 billion years ago, said J. D. A. Piper at Oliver Lodge Laboratory in Liverpool, and to the Grenville Belts of 1.15 billion years ago. Two other events, said Piper, the Katangan of 680-580 million years ago (mya), and the Damarian of 550-500 mya, are also broadly coincident with hairpin turns.

Same in the Carboniferous. In the upper Carboniferous, said the Canadian scientists Irving and Robertson, a rapid change in plate movement direction occurred immediately before the onset of persistent reversed field.

A hairpin turn correlates with the Appalachian orogeny and with the opening of the Laurasian Atlantic, said Irving, a hairpin turn about 1.8 billion years ago must "surely" relate to the Hudsonian Orogeny, and there's the "exciting possibility" that a simultaneous change in polar wander, geomagnetic variations, and plate movement direction took place 24 mya at the end-Oligocene.

No one knows why hairpin turns occur, but they appear to have been aborted polarity reversals. (Which means that tectonic activity not only coincides with reversals, it also coincides with aborted reversals).

And that's how Mother Nature hid her tracks. That's why we never noticed the cycle; we didn't include aborted reversals in our calculations. But there *is* a cycle, a cycle that includes orogenesis, seismic activity, sea level changes, black shale deposition, volcanism, extinctions, seafloor spreading, and magnetic reversals.

To see the cycle, take another look at the sea level chart by P. R. Vail (next page). Because sea level declines occurred so abruptly, they've received most of our attention. Look again, however, and you'll see that many sea level rises began abruptly too.

If sea level rises coincide with tectonic movement, shouldn't each upward bump on the chart be as important as each drop? Draw a line where each rise began, another at each drop, and a startling 14.1-million-year cycle pops right off the chart.

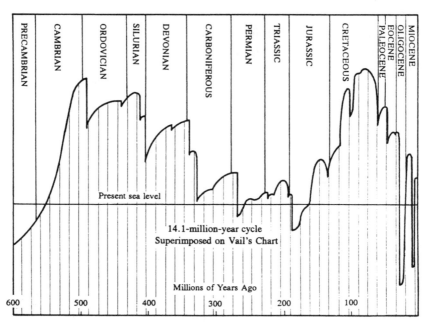

14.1-million-year cycle superimposed on Vail's chart.

Of the 25 major sea level changes in the past 600 million years, the 14.1-million-year cycle lines up almost exactly with 24. Such a success rate (96%) is unheard of. Any process that repeats itself time after time for 600 million years didn't happen just by chance.

Look, too, at how the cycle corresponds to geological periods. Any pattern that falls almost exactly at the ends of eight out of ten geological periods and comes within spitting distance of the other two, isn't just a throw of the dice. And there's more. The cycle also corresponds with the end-Eocene and end-Oligocene.

But wait until you see all of the other things that have happened on our planet every 14.1 million years—bearing in mind that two to three million years one way or the other is considered "close enough" by most paleontologists.

10 mya - Mid-Miocene extinction in the seas. Extensive glaciation (Crowley and North). Abrupt increase in underwater volcanism (Heath). Iridium spike (Raup). Rapid geomagnetic intensity fluctuations (Prévot).

24 mya - End-Oligocene extinction. Major underwater volcanism. Major volcanism on land (Axelrod). Change in direction of seafloor spreading. Magnetic reversal (Kent and Gradstein).

37 mya - End-Eocene extinction. Sea level drop. Major explosive non-marine volcanism (Axelrod). Huge flood basalts in Ethiopia. Rapid glaciation (Shackleton). Change in direction of seafloor spreading. Major underwater volcanism. Tectonic episodes (Rampino and Stothers). Microtektites. Iridium spike (Raup). Peak in magnetic reversal frequency. *(Warning! We're going through a period of frequent magnetic reversals right now!)* (Merrill)

50 mya - Natural oil spill in Cuba (Kurtén). The Ancestral Rocky Mountains blew apart in a matter of hours (Hsü). Sea level drop. Nova Scotia's Montagnais Crater formed (Raup). Magnetic reversal (Kent and Gradstein).

65 mya - K-T extinction. Microtektites. Worldwide volcanism (ash is more than three feet deep in places). The Deccan Traps spewed out more than two million cubic kilometers of basalt. The Brito-Arctic flows spewed out more than 500,000 cubic km of basalt. Mountain building. Sea level drop. Popigai, Manson, and Chicxulub Craters. Worldwide cooling (but temperatures in the seas went up). Massive underwater volcanism. Black shales. Iridium spike and many other elements. Magnetic reversal.

79 mya - Tectonic episodes (Rampino and Stothers). Massive rift volcanism smothers more than 30 million square kilometers (Rea and Vallier). Change in direction of seafloor spreading (Atwater). Jack Horner's herd of 10,000 dinosaurs bites the bullet (more later). Extinction. Magnetic reversal (Vail).

93 mya - End-Cenomanian extinction. Iridium spike (Raup). Non-marine volcanism. Black shales (Hallam). Major underwater volcanism (the Pacific Superswell). Glaciation (Crowley and North).

108 mya - Aptian extinction. Sea level rise. India's Rajmahal Basalt flows spewed out 200,000 cubic km of basalt. Intense terrestrial volcanism (Axelrod). Major underwater volcanism (Schlanger *et al*). Black shales (Hallam). Glaciation (Hallam). Magnetic reversal (Vail).

120 mya - Mid-plate volcanism (Schlanger, Jenkyns, and Premoli-Silva). Magnetic reversal (Gradstein *et al*).

144 mya - End-Jurassic extinction. Rapid seafloor spreading (Sheridan). Magnetic reversal (Vail). Black shales (Hallam). Climatic cooling (G. Warren). Major uplift and orogeny (Hallam). Two million cubic km of basalt spewed from the Paraná flows to cover parts of Paraguay, Uruguay, Brazil and northern Argentina. Another 500,000 cubic kilometers poured from the Namibian basalts in Africa.

163 mya - End-Callovian extinction. Sea level rise. Rapid seafloor spreading (Sheridan). Microtektites. Magnetic reversal (Vail). Iridium spike (Raup).

177 mya - Bajocian extinction. Magnetic reversal (Vail). Glaciation (Erickson). Major flood basalts from a 2,400-mile-long rift in the Antarctic and in Tasmania. Sea level drop.

190 mya - Pliensbachian extinction. Black shales (Hallam). Tectonic episodes (Rampino and Stothers). Karoo Flood Basalts in South Africa pumped out two million cubic km of basalt. Rapid polarity shifts (Vail).

206 mya - Rhaetian/Norian extinction. Spikes in iridium, platinum, osmium, chromium, other elements (Hallam). Major seafloor spreading. Flood basalts cover more than 100,000 square km of eastern North America from Nova Scotia to North Carolina. Black shales (Hallam). Frequent magnetic reversals (Creer). Glaciation (Fabricius, Friedrichsen and Jacobshagen). Canada's Manicougan Crater forms. Magnetic reversal (Gradstein *et al*).

248 mya - The Great Permian extinction.   Major underwater vol-
          canism (Courtillot). The Siberian Traps. More than 1.5
          million cubic km of basalt up to 2.32 miles deep smoth-
          ered more than 130,000 square miles of Siberia. Stron-
          tium spike (Hallam). Iridium spike, and many other ele-
          ments (Clark *et al*). Carbon spike (Erwin). Climatic
          cooling (Stanley). Black shales (Hallam). Magnetic
          reversal. (Gradstein *et al.*)

260 mya - Tectonic episodes (Rampino and Stothers).

286 mya - End-Carboniferous extinction. Rapid changes in polar
          wander. Wide-spread uplift (Hallam). Major glacia-
          tion (Stanley).

320 mya - End-Mississippian extinction. Spikes in iridium, plati-
          num, osmium, chromium, and other elements (Hallam).
          Major glaciation (Hallam).

330 mya - Glaciation. Black shales (Hallam). Extinction (Crow-
          ley and North).

345 mya - Tectonic episodes (Rampino and Stothers).

370 mya - Frasnian/Famennian extinction. Black shales (Digby
          McLaren). Tectonic episodes (Rampino and Stothers).
          Climatic cooling (Stanley). Iridium spike (Thomas).

408 mya - End-Silurian. Black shales. Sea level drop. Major gla-
          ciation (Steiner and Grillmair).

444 mya - End-Ordovician extinction. Glaciation into the middle
          of the Sahara. Major seafloor spreading. Black shales
          (Hallam). Strontium spike (Hallam). Frequent magnetic
          reversals (Hays). Iridium spike (Donovan).

505 mya - End-Cambrian extinction. Tectonic episodes. Glacia-
          tion. Black shales (Stanley). Frequent magnetic re-
          versals (Creer).

558 mya - Extinction.

590 mya - Extinction. Cambrian explosion. Glaciation (Ward).

657 mya - Extinction. Greatest glaciation in history (Stanley).

699 mya - Extinction. Widespread glaciation.

The proof lies in the numbers. All of those things occurred in almost perfect harmony with the 14.1-myr cycle. (Did you notice how many times glaciation or cooling was mentioned?)

Craters match the cycle, too. Most "meteor" craters formed 28.4 million years apart [at every other pulse of the 14.1-million-year cycle], said Walter Alvarez and Richard Muller, after studying a list of craters compiled by the Canadian scientist R. A. F. Grieve. Not only did the craters form about 28.4 million years apart, they also formed in phase with extinctions. The probability that such an agreement is accidental, said Alvarez (that crater formation would match the extinction cycle), is one in a thousand.

Alvarez had one little problem, though. The eight-mile-wide Lappäjarvi Crater in Finland, which dated at 78 $\pm$ 2 million years ago, refused to fit the 28.4-million-year cycle. But Lappäjarvi—*which formed at a reversal*—fits the 14.1-million-year cycle to perfection! What odds would Alvarez give now, I wonder? One in a million?

Other craters formed at reversals, too. The Ries crater formed at a reversal about 15 million years ago (Glass, Swinckl, and Zwart), while Ghana's Bosumtri crater formed at a reversal about two million years ago at the end-Pliocene. Another crater, the 128-million-year-old Tookoonooka Crater in Queensland, Australia, also formed near a reversal. (I think most so-called "meteor" craters were formed by massive underground explosions triggered by magnetic reversals. But that's another story.)

Grieve's own formulas give another boost to the cycle. On average, said Grieve, a one kilometer crater is formed every 1,400 years, a 10 kilometer crater is formed every 140,000 years, and a 100 kilometer crater is formed every 14 million years.

How much more plain could it be?

There's a 14.1-million-year cycle to large craters; there's a 14.1-million-year cycle to tectonic movement; there's a 14.1-million-year cycle to sea level changes; *and there's a 14.1-million-year cycle to magnetic reversals.*

This is where the French scientist Alain Mazaud of the University of Paris comes into the picture. Not only is there a 30-million-year pattern to magnetic reversals, said Mazaud, there's also a less pronounced cycle at 15 million years (give or take two million). The reversals alternated in strength, with every other one a bit stronger and falling approximately at an extinction.

Mazaud is adamant. "Regardless of its origin, a periodicity of 15 million years [give or take two million] . . . represents one of the most persistent periodic, long-term geophysical phenomenon [*sic*] ever reported." (It happens, I suspect, each time we bob through the galactic plane and again as we reach the furthest point away from the plane where we reverse direction and head the other way.)

With a 14.1-million-year cycle to every kind of tectonic activity there is, and now, a 14.1-million-year cycle to magnetic reversals, how can we possibly pretend that they aren't related?

Now, let's see how tectonic forces—especially underwater volcanism—teamed up with those reversals to cause havoc and death in the seas . . . and to cause ice ages.

There is evidence that climate, volcanism, tectonic activity, cratering, and magnetic reversals may all be correlated.

—DAVID LOPER

10

. . . . . .

FISH STEW

. . . . . .

Underwater volcanoes affect our lives and our weather in ways we don't understand. But how can we understand, when we don't know how many there are? How can we understand, when we keep finding new ones by the thousands?

Marine geophysicists aboard the research vessel *Melville* recently discovered 1,133 previously unmapped underwater volcanoes about

600 miles northwest of Easter Island. (Easter Island is about 2,300 miles west of Chile in the South Pacific.)

And they're huge. Some of the newly-found volcanoes rise almost a mile and a half above the seafloor. Even then, their peaks remain about a mile and a half below the water's surface. Consisting of both seamounts and volcanic cones, they're packed into an area of 55,000 square miles, about the size of New York state.

Scientists are shocked. "We thought we would find a few dozen new volcanoes," said Ken Macdonald at the University of California, Santa Barbara. "Instead we found over 1,000 that had never been mapped before." It's the greatest concentration of geologically active volcanoes on earth. (*Seattle Times,* 14 Feb 1993)

We have no idea how many volcanoes may be lurking beneath the seas. Oceanographers previously believed there were about 10,000. But now? It's up for grabs. Only five percent of the ocean floor, said Macdonald, has ever been mapped in detail.

What we do know, is that underwater volcanoes pump awesome amounts of heat into the seas. The area near the East Pacific Rise where the new volcanoes were discovered is a region of intense volcanic activity where red-hot basalt gushes from the earth's fiery interior and plumes of water heated to almost 800 degrees Fahrenheit spout into the inky blackness.

Such large injections of heat may change seawater temperatures enough, said geophysicist Dr. Daniel A. Walker of the University of Hawaii, to affect water circulation patterns, and may even trigger El Niño.

Sometimes called "the earth's heartbeat," El Niño is a weather phenomenon that usually occurs every three to seven years when a massive high-pressure system near Easter Island loses its punch.[1] It's caused by a 450-foot-thick slab of water heated to about 14°F above normal. Stretching some 8,000 miles along the equator, this wedge of superheated water covers an area about equal in size to the United States.

---

1. El Niño, the Child, alludes in Spanish to the Christ child, because its warm currents often reach South America about Christmastime.

As this enormous expanse of warm water evaporates (the warmer the water, the more evaporation), it forms huge rain clouds. At the same time, Pacific trade winds, which normally flow east to west, fade or even switch directions. Higher up, the jet stream splits in two, altering weather systems around the globe.

El Niño changes rainfall patterns as far north as Seattle and much of the Pacific Northwest. It causes torrential rainstorms in parts of the world up to twenty times normal, brings famine to Africa, floods to California, and droughts to Australia.

During the 1982-'83 El Niño, for example, an incredible 100 inches of rain fell on some parts of Ecuador in six months, creating inland lakes where previously there had been only desert. It cut a swath of destruction across five continents, caused drought in India, dust storms and brushfires in Australia, and drove an estimated 17 million nesting birds from their island homes in the Pacific. Worldwide, the so-called El Niño of the Century caused about 8.11 billion dollars damage.

Embarrassingly, the 1982 El Niño was in full swing before climatologists even realized what was happening . . . or its importance. They now realize that El Niño may be the largest single weather influence on the planet. "El Niño is the most robust cause of global climate variability we've ever identified," says oceanographer Michael McPhaden of the National Oceanic and Atmospheric Administration (NOAA).

And its influence is growing. During the last decade El Niños have become both longer-lasting and more frequent.

The future looks even bleaker. Columbia's Mark Cane and Steven Szebiak, "the godfathers of computer weather modeling," predict more severe—and more frequent—El Niños during the next century. (From "El Niño the Weathermaker," Douglas Gantenbein, *Popular Science*, May 1995)

In the past, ocean temperatures have see-sawed above or below normal every three to seven years. But "that pattern fell apart in the mid 1970s," said science writer Richard Monastersky, "when the Pacific's equatorial thermostat stuck on warm. Since then, five El Niños have brewed, but only once did the ocean cool off." (From

"Tropical Trouble: Two decades of Pacific warmth have fired up the globe," *Science News*, 11 Mar 1995)

Some scientists dismiss the El Niño/underwater volcanism theory as absurd. But it was given a strong boost by oceanographer Dr. D. James Baker, administrator of the National Oceanic and Atmospheric Administration, who highlighted it at a recent symposium as a possible explanation. New research, partly funded by NOAA, reveals that underwater volcanism is "one of the strongest and least understood forces on earth, producing a vast heating of the seas equal to that of 3,000 big nuclear reactors." (From "Hot Vents in the Sea Floor May Drive El Niño," *New York Times*, 25 Apr 1995, by William Broad.)

In addition to those near Easter Island, unmapped underwater volcanoes have also been found in the north. An active underwater volcano was discovered in June 1993 about 270 miles west of Astoria, Oregon. Named Coaxial Volcano, it's a four-mile-long gash in the seafloor lying beneath a mile and a half of water on the Juan de Fuca Ridge. Accompanied by swarms of earthquakes, hot water spews from one end as molten lava gushes from the other. (Bellevue, Washington, *Journal American,* 20 Aug 1993)

And in early 1996, oceanographers discovered yet another monster underwater furnace about 125 miles south of Coaxial Volcano off the coast of Newport, Oregon. This fiery groove in the ocean floor is six miles long, some 50% longer than its northern neighbor. (KING 5 TV, Seattle, 14 Mar 1996)

This is going on right now, folks!

Eight-hundred-degree plumes of water, from generally small lava flows, are gushing into our seas right now, from the coast of Oregon to the South Pacific.

And they're heating our seas!

A recent study at the Monterey Bay Aquarium Research Institute in Pacific Grove, California, showed that seawater temperatures are rising. Water temperatures in Monterey Bay have increased almost one degree Centigrade during the last 60 years alone.

The increase is causing profound changes in underwater populations as certain species of starfish, snails, crabs, and anemone follow the warm waters north. Significant population increases have occurred in eight species more common to southern California, while the abundance of five northern species has declined. (Barry, Baxter, Sagarin, and Gilman, *Science*, 3 Feb 1995)

This increase in sea temperature, let me repeat, is from "generally small" lava flows.

Imagine how hot it must have been above the massive underwater lava flows of the end-Cretaceous! (Remember, according to Maurice Ewing, all 46,600 miles of our present-day underwater ridge system may have been initiated at the time.)

Staggering amounts of basalt must have poured into the seas!

The numbers are astronomical. Some 80% of all volcanic eruptions occur underwater, said Steve Hammond, manager of NOAA's Vents Program at the Hatfield Marine Science Center in Newport, Oregon. (Bellevue, Washington, *Journal American*, 18 Aug 1993)

Using that 80% ratio, and knowing that two and a half million cubic kilometers of basalt spewed out of the Deccan Traps and Brito-Arctic Flows alone, up to 10 million cubic kilometers (three million cubic miles) of basalt could have sizzled into end-Cretaceous seas almost overnight.[1]

Ten million cubic kilometers of basalt would smother every state in the United States, including Alaska and Hawaii, under a solid layer of rock more than one kilometer (6/10ths of a mile) thick.

And every one of those ten million cubic kilometers of basalt would have been unbelievably hot. Up to 2,150°F hot. A planetary-sized hot water heater. That's why ocean temperatures soared.

The seas must have boiled—literally boiled—above the underwater ridges!

Fish stew. Clam chowder. Bouillabaisse. Shells and all. The biggest stew pot in the world. No wonder so many fish went belly up. That's why deep-water animals fared better; it's cooler down there.

---

1. Ten million cubic kilometers would be only a small percentage of today's ridge system. The volume of existing ridges is roughly 1.6×10^8 km^3 (160,000,000 cubic km). (Peter A. Rona, *Geol. Soc. Am. Bull. V. 84.*)

We *know* the oceans can boil. In 1963, as hot magma spewed from the Mid-Atlantic Ridge during the birth of the Icelandic island of Surtsey, the ocean boiled, and steam rose into the sky. "Its enormous output of steam, if harnessed," said volcanologist and eyewitness Sigurdur Thórarinsson, "would surpass the energy produced by Niagara Falls 100,000 times."

Same in Hawaii. As lava flows into the sea, said Herbert Shaw and James Moore of the USGS, it creates lines of steaming water to mark its progress. We assume, they said, that the seas could actually boil above shallow mid-ocean spreading centers. (*Eos*, 8 Nov 1988)

Pity any fish in those superheated seas . . . instant extinction.

Indeed, underwater volcanism and extinctions do go hand-in-hand. There is a "remarkable correspondence," said Rich, Johnson, Jones, and Campsie in 1986, "between variations in seafloor spreading and marine extinction." "Hydrothermal activity [heated water] . . . may provide a partial explanation."

But still, warmer seas don't explain it all. Think back to the summer days of your youth, when the air was hot and humid and your shirt clung to your back as if it were painted on; days when you'd have given your eye-teeth for a sparkling pool of water; days when huge thunderstorms rumbled across leaden skies and jagged flashes of lightning leapt through the clouds.

Remember your mama's words? Don't go swimming in a thunderstorm, she warned, the lightning could kill you. Electricity will shoot through the water and you'll die. Well, mama was right.

And that's what happened at the K-T. If comparatively small volcanoes such as Surtsey, Mount St. Helens, and Westdahl can generate powerful electrical storms, wouldn't it seem that underwater volcanoes big enough to pump millions of cubic kilometers of basalt into the sea should generate enormous amounts of electricity?

Billions upon billions of volts of electricity must have raced through end-Cretaceous seas, plowing into every underwater nook and cranny on earth.

It's possible. Have you seen those clothes-wrinkle removers in the stores lately? Made of plastic, they look like ordinary steam irons.

But you don't use them to press your clothes, you use them to steam your clothes. Run by electrolysis, they shoot an electric current through water, heat it to the boiling point, and generate oodles of steam—much more steam than an ordinary steam iron. But there's a catch. They don't work unless you add a pinch of salt.

Try it, you can buy one for about twenty bucks. First, use ordinary tap water. Nothing will happen. Plug it in for an hour, a day, or a week . . . nothing. Now try again, but this time add the salt. Electricity will race through the water like a rumor through a crowd and your steamer will be boiling in seconds. But you've got to add the salt.

The same thing happened in end-Cretaceous seas, I submit, when the earth's polarity flipped. Electricity shot through the water at the speed of light, racing the 24,000 miles around the equator in less than one second, less than the snap of a finger—and ZAP!—like shoving an electric cord into a hot tub, instant death for millions of fish.

Then it stopped.

Stopped at the mouths of rivers and streams where freshwater blocked the electrolysis. The seas contain salt, so they conducted electricity. Freshwater doesn't, so it didn't. It's one more reason why sea-dwellers died and freshwater animals survived, one more reason why bizarre life-forms appear immediately after extinctions. Ionized water (charged water), say geneticists, can cause mutations.

Electrolysis or not, though, we know that thousands, perhaps millions, of cubic miles of red-hot basalt spewed into end-Cretaceous seas, initiating our present-day underwater ridge system. We also know the seas had spread far inland, creating hundreds of thousands of additional square miles of ocean surface, and that ocean temperatures had risen by as much as 22°F. Combine the two, and evaporation had to have increased dramatically.

Unbelievable amounts of moisture had to have risen into skies already clogged with volcanic debris; skies therefore dark and frigid.

Warmer seas and colder skies . . . a deadly combination.

What happens to steam when it cools? It condenses. That moisture had to have condensed and fallen to the ground. Rains of Biblical proportions had to have pounded on the newly risen mountains and eroded them, sending torrents of mud and water into swollen rivers and streams and back to the thirsty seas.

And flood it did. That's how dinosaur bone quarries such as the Red Deer River in Alberta, and Dinosaur National Monument in Utah, were created. Dinosaur bones are "stacked like cordwood in a logjam" in those massive dinosaur graveyards, sometimes more than ten stories deep. How do you get ten-story stacks of bones? From ten-story floods, of course.

Further proof of huge floods comes from the low oxygen-18 levels found in earliest Tertiary fossils. Low oxygen-18 levels are important because freshwater is deficient in heavier oxygen isotopes. Finding low oxygen-18 levels in fossils, said Hans Thierstein and Wolf Berger of Scripps Institution of Oceanography, means that end-Cretaceous seas must have received "a large injection of fresh water."

That's why huge gaps appear in the geologic column at extinctions. Gigantic floods eroded the missing soil. Take the Great Permian. Almost three-million-year's worth of sedimentary rocks are missing at the Great Permian extinction, said David M. Raup, in his 1991 book *Extinction: Bad Genes or Bad Luck?*

Changes in strontium ratios indicate flooding, too. The $^{87}Sr/\,^{86}Sr$ seawater ratio increased gradually during the late Cretaceous and into the early Tertiary, said Officer, Drake, Hallam, and Devine, but was interrupted by a sharp rise and fall right at the boundary. That spike in the ratio, they believe, indicates an increase in continental run-off, which "implies flooding."

Anyway you slice it, unbelievable amounts of moisture had to have fallen from those ash-filled K-T skies.

Now for the biggest question of all. What would have happened to all of that precipitation if it had been cold at the time?

Snow. Unimaginable amounts of snow.

The possibility that there is some connection between [magnetic] reversals, climatic changes, and extinctions cannot be ruled out.
—JAMES D. HAYS

11

. . . .

NOT BY FIRE BUT BY ICE

. . . .

A long time ago, the universe was made of ice. Then one day the ice began to melt, and a mist rose into the sky. Out of the mist came a giant made of frost, and the earth and the heavens were made from his body. That is how the world began, and that is how the world will end, not by fire but by ice. The seas will freeze, and winters will never end.
—ANCIENT SCANDINAVIAN LEGEND

Those ancient Scandinavians had it right, their world began by ice, and that's how it will end. That's how the dinosaurs' world ended, too . . . by ice.

Temperatures plummeted from an average of 20°C (68°F) to an average of 10°C (50°F) at the end-Cretaceous, said Jack Wolfe of the USGS. It was cold—and it was wet. Put cold and wet together and what do you get?

Snowstorms from hell. "There could have been 200 inches of precipitation a day for weeks," said Kenneth Hsü, in his 1986 book *The Great Dying*. "Recalling the drop in temperature during just that period, it would have fallen not as rain, but as snow. The globe would have been a giant snowball when the sun shone again."

Wow! Two hundred inches of rain turned to snow! Per day!

How much snow would that be? Just add a zero. One inch of rain, say meteorologists, translates into ten inches of snow. Two hundred inches of rain translates into an incredible 2,000 inches of snow.

That's 166 feet! Sixteen stories! A day! For weeks! No wonder sea levels fell! What good would it be to be five stories tall if sixteen stories of snow fell on your head? Every day? For weeks?

The dinosaurs' world ended by ice.

And that's how our own world will end . . . by ice.

Ice played a major role at other extinctions, too. Climatic cooling was the "dominant agent" at the Cambrian extinction, said Steven M. Stanley in his 1987 book *Extinction,* as it was at the Ordovician, the Permian, the Devonian, the Carboniferous, the mid-Miocene, and yet again at the end-Miocene.

Indeed, Stanley thinks climatic cooling caused almost every extinction on record. The end-Triassic, end-Jurassic, end-Silurian, and end-Eocene were times of glaciation, and four extinctions during the Cambrian, says paleontologist Rob Thomas of West Montana College, can also be attributed to glaciation. The Precambrian extinction saw extensive glaciation, as did an extinction about 657 million years ago, when thick layers of ice spread into Australia, further south than at any other time in history.

Now . . . take a wild guess as to what those glaciations had in common.

Magnetic reversals.

The end-Carboniferous, end-Ordovician, end-Permian, end-Cambrian, and end-Triassic were all times of frequent magnetic reversals or changes in polar wander, while the Cretaceous, Eocene and Miocene actually ended with magnetic reversals.

At least twelve reversals can be linked to extinctions and climatic deterioration during the last three million years alone. A reversal about three million years ago marked the onset of glaciation, said James D. Hays of Lamont-Doherty Earth Observatory. A reversal about two million years ago at the end-Pliocene marked the onset of glaciation. (The end-Pliocene also showed a dramatic increase in volcanism.) A reversal about one million years ago marked the onset of glaciation, and so did the Brunhes reversal of 780,000 years ago. "The evidence suggests," said Hays, "that environmental change [is] connected in some way with the reversal of the earth's magnetic field." The Jaramillo reversal of 850,000 years ago also saw glaciation (Ardrey).

The Gothenburg excursion of 11,500 years ago can be linked to glaciation, as can the Lake Mungo and Mono Lake excursions. The Blake magnetic reversal of 115,000 years ago correlates with glaciation, and so do the Biwa I, Biwa II, and Biwa III reversals. (These reversals are discussed more thoroughly in Chapter 16.)

Why do climatic changes coincide with magnetic reversals? Because in normal times, I believe, our magnetic field holds tectonic forces in check. But when the field weakens during a reversal, that balance disappears. Suddenly unleashed, underwater volcanoes heat the seas and excess moisture rises into the sky. Then the moisture condenses and falls to the earth as giant snowstorms and giant floods. Warmer seas, said W. F. Ruddiman, of Lamont-Doherty Earth Observatory, provide the optimum configuration for rapid glaciation.

And that's why ice ages correlate with volcanism.

A similar theory, but without reference to magnetic reversals, came from Dr. L. J. Krige, president of the Geological Society of South Africa. We have long known, said Krige, that ice ages accompany or follow periods of mountain building. We also know that magmatic events occurred at the same time. Since most tillite deposition corresponds to the same cycle, he said, ice ages may be caused by underwater volcanism.

Magmatic activity in the seas, Krige explained, would promote so much evaporation that rainfall would increase to 1,500 centimeters (580 inches) per year. Since the accompanying increased cloudiness would cause decreasing temperatures, most of the precipitation would fall as snow. "The inevitable consequence," said Krige, "would be a continual increase in the amount of snow . . . in other words, an ice age." ("Magmatic Cycles, Continental Drift and Ice Ages," *Proceedings of the Geological Society of South Africa*, 1929)

That was 1929. Who would have dreamed that Krige's underwater volcanism/ice age theory would one day be combined with Motonari Matuyama's magnetic reversal theory, presented just a few months earlier?

Who would have dreamed, for that matter, that much of our planet had once been covered by ice? Less than 200 years ago, though huge erratic boulders, some as big as a house, were strewn over Europe, we didn't have a clue that an ice age, an *eizeit*, had ever occurred.

The boulders were transported to their present locations by Noah's Deluge, said scientists of the day. But the young Swiss naturalist Louis Agassiz disagreed. There is no way, thought Agassiz, that floods could account for the scrapes and grooves that he saw everywhere in the bedrock. And there is no way that floods, no matter how big, could explain the huge boulders.

Though wrong, the official explanation for the boulders at least flirted with the truth; it at least associated them with ice. The boulders had been trapped in icebergs, said Charles Lyell in his 1833 book *Principles of Geology*, then drifted to their present locations when the seas were much deeper than today.

That sounded logical. Sailors of the day had seen rocks frozen in icebergs. Even Charles Darwin believed the floating rock story. He'd seen rocks frozen in icebergs, too.

But some erratic boulders lie stranded more than a mile above the nearest ocean. No one could believe the seas had been that much deeper. Impossible. Where did the water come from? Where did it go when the flood ended?

Waters a mile deep poured upward from hidden underground reservoirs, yelled frantic flood-believers. Thousands of feet of water simply—appeared. It gushed out of solid ground, flooded the entire world, and just as quickly—disappeared. Where did it go? The world is a sponge, they swore. It sank into secret hidden caverns; into uncharted, unfindable, unexplainable caverns. Just have faith.

Well, then, maybe the earth wobbled on its axis, and tidal waves washed over the tops of the highest mountains. Or maybe (they were getting desperate now) a huge comet skimmed across the world's oceans like a skipping-stone on the old millpond, and mountain-high waves rippled away from each point of impact in ever wider circles.

Sure, blame it on a comet. That's what we're doing right now, isn't it?

Convenient culprits, those comets.

No, don't blame a comet. Blame ice. Maybe we don't know where the dinosaur's seawater went, but we know *exactly* where the mammoth's water went. Since the mammoths went extinct a mere 11,500 years ago, the clues are still fresh. Their water turned to ice and got locked up on land. That's why sea levels at the end of the last ice age were 350 to 400 feet lower than today.

Though we've been in a warming trend for more than 10,000 years, close to seven million cubic miles of ice up to one mile deep still smother Antarctica. That's enough ice to cover the continental United States almost two miles deep. More than another 500,000 cubic miles of ice cover much of Greenland and other parts of the north. Put it all together, and close to 10% of the world's landmass is covered by ice.

Spread it out, and you'd have a 40-story-thick layer of ice across every continent on earth.

If all that ice were to melt, sea levels would rise 65 to 80 meters (214 to 264 feet). Los Angeles, New York City, London, Rio de Janeiro, Hong Kong, Boston, hundreds of cities around the globe, would disappear under the rising frigid waters. Shorelines would move inland about 70 miles in most places.

Our present ice age (or the most recent one, depending on how you look at it), began about 115,000 years ago. During the next 5,000 years the ice sheets grew an incredible 1,200 cubic miles per year, then kept growing (but more slowly) for another 6,000 years. During the next 11,500 years, they pulled back. Then they advanced. Back and forth they went every 11,500 years, until, reaching a peak some 60,000 years ago, massive sheets of ice sprawled across most of Canada and Scandinavia.

Approximate extent of glaciation today (above) and during the last ice age (right). Twenty-thousand years ago, great ice sheets covered parts of North America, Europe, and Asia. Surface waters of the Arctic and parts of the North Atlantic Ocean were frozen, and sea levels were 350 feet lower than today. Many parts of the continental shelf, including a "bridge" between Asia and North America, became dry land.

Both drawings by Anastasia Sotiropoulos, based on information compiled by George Denton and other members of CLIMAP (Climate: Long-range Investigation Mapping and Prediction), as originally published in John and Katherine Imbrie's book *Ice Ages: Solving the Mystery*. Reprinted by permission of Enslow Publishers.

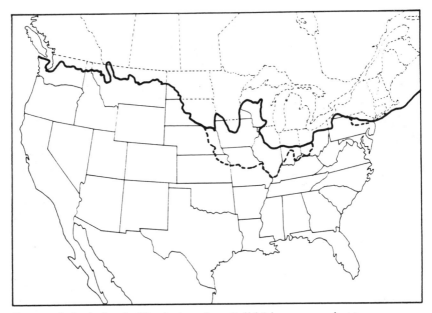

Extent of glaciation in North America. Solid Line: approximate
extent of glaciation during last (Wisconsin) ice age. Dotted line:
approximate extent of previous glaciations.

Then they retreated. But not for long. By 35,000 years ago, they
again crept down the mountain flanks, only to pull back, then advance yet again about 23,000 years ago. Millions of cubic miles of
water had turned to ice measuring one to two miles deep, and
covered a third of the surface of the earth.

How do glaciers form? When snow gets to be about 100 feet deep,
its sheer weight compresses it into ice. Then it turns into glaciers.
Hungry, sluggish monsters, the advancing glaciers eat everything in
their paths, swallowing huge chunks of solid rock as if they were
candy. Then they use the rocks to grind on the underlying bedrock,
sometimes smoothing it, sometimes polishing it, sometimes gouging
and scratching it.

Glaciers more than 10,000 feet deep flowed southward out of
Scandinavia and Scotland, burying forests, fields, rivers, and mountains. Like an advancing army, they extended their icy grips far

south into parts of England, Germany, Denmark, Poland, and all of Ireland. Polar bears and reindeer roamed the outskirts of England and central Europe.

In southern Europe, ice filled the valleys to their very tops until rivers of ice spewed between the peaks and tumbled down the mountains. Pouring out of the Pyrenees, the Alps, the Apennines, and the Carpathians, even from Africa's Mount Kilimanjaro, they buried parts of France, Italy, Austria, and all of Switzerland.

In North America the ice marched southward from Hudson Bay to obliterate all of eastern Canada. Again, it reached depths of one to two miles. We know its depth by looking at mountains, such as Mt. Monadnock in New Hampshire, which were partly submerged. There's a line on the mountains where the ice ended. Below the line, the mountains are smooth and even. Above it, rough and craggy.

Ice moved out of the Canadian Rockies like cavalry from the fort, attacking entire sections of Alaska and most of western Canada. Creeping southward out of British Columbia, it crossed the border into Idaho, Montana, and Washington. Scraping the flanks of the Olympic Range to the west and the high Cascades to the east, it lumbered south through Washington state to Olympia, burying Seattle under 3,000 to 5,000 feet of ice on its regal procession to the capital.

Moving on several fronts at once, the icy killer attacked Illinois, Indiana, Ohio, Iowa, and much of the Midwest, moving far south of Chicago to the Ohio and Missouri Rivers. Bulldozing its way into New England, it rolled across the state of Maine, past the coast, and to the very edge of the continental shelf.

Past what is now the coast, that is. Sea levels were so low that the entire continental shelf, at least in eastern North America, was above water. Many states on the eastern seaboard were twice as big as today. New Jersey's shoreline, for example, was 60 to 100 miles east of its present location.

Same in the west.

The land between Alaska and Asia rose out of the sea like a bridge (or rather, the sea dropped away from the land), and the Bering Strait, which today is only 18 stories deep at its deepest point, was above water. You could have *walked* to Alaska. (The word *bridge* is misleading. The land connection between Alaska and Siberia was almost as wide as Alaska itself.)

Same all over the world. England was connected to France, Australia was connected to Tasmania and New Guinea, Sumatra was connected to the mainland, Japan was connected to Korea, Sri Lanka was connected to India, and Venice lay 200 miles from the sea. And this is only a partial list.

When the glaciers ended, they ended in a cliff towering as much as 15 stories above the surrounding land, an ice-cliff stretching from eastern Long Island to Washington state. And for some reason, the ice had a northern boundary, too. A relatively thin layer of floating ice about the same as today covered the Arctic Ocean.

By the time it stopped, the cold-hearted invader from the north had claimed almost three times as much land as today: more than 17 million square miles. Ten million square miles of the northern hemisphere, 13 times more than now, was covered by ice. The southern hemisphere had only slightly more than at present.

You'd think an onslaught such as that would have affected the entire planet, but amazingly, the rest of the world barely changed. The tropics and subtropics were only four degrees colder, while the equatorial rainforest belt remained much the same as today.

If we had that much ice, and temperatures were only four degrees colder, imagine how much ice we must have had at the K-T, when temperatures were 18 degrees colder and the seas had dropped a fifth of a mile!

Then, about 11,500 years ago, the ice age ended. And it ended *fast*. As the world grew warmer, tropical animals moved back into Europe, and the barren tundra filled with trees once again.

Further south, where there had been no ice to begin with, rainfall patterns changed. Entire forests withered and died away as parts of the world that had been cool and wet, such as the western United States, suddenly turned hot and dry.

And the woolly mammoth disappeared. It was a global sweep of death—a mass extinction—destroying not only the mammoth, but some 75% of all of America's larger mammals.

But why only the *big* ones? And why so fast?

12

. . . .

FATAL FLAW

. . . .

Fat, happy, and healthy, there was no reason for the woolly mammoth to die. Moving south each time the ice advanced, then pulling back to the harsher weather it preferred each time the ice retreated, it had tiptoed through the tundra for two million years, dancing its way through at least four previous ice ages. Then, about 11,500 years ago, our most recent ice age came to an end . . . and so did the woolly mammoth. No one knows why.

Whatever the reason, it took 40 million of the world's mammals on the same one-way walk to oblivion. And, as with the dinosaur extinction of 65 million years ago, the killer zeroed in on the big boys, on species where the adults weighed at least 100 pounds.

Big was bad.

Seventy-two percent of all large animal species went extinct, said David M. Raup in his 1991 book *Extinction: Bad Genes or Bad Luck?* whereas only 10% of small species disappeared. "The preponderance of extinctions among large mammals," said Raup, "is not likely to be due merely to chance. Large size really did put land mammals at a much higher risk of extinction."

Big animals were hammered; small ones merely tapped on the shoulder. (Except on islands. For some reason, small mammals and birds on oceanic islands were severely affected.) The mastodon died, the woolly rhinoceros died, so did the saber toothed cat and the toxodont. (The saber toothed "cat" was as big as today's lion, while the toxodont, built like a low-slung rhinoceros, stood five feet tall at the shoulder and measured up to nine feet long.) Did every one of those big bulky bruisers have the same fatal flaw?

And it wasn't the only time. Big land mammals were also extremely hard hit about 37 million years ago at the end-Eocene, said paleontologist Robert Bakker, as they were some two million years ago at the end-Pliocene when about 70% of the world's larger mammals went extinct. Same at the end-Miocene, when almost two-thirds of the 62 genera that went extinct were of big body size. (Steven M. Stanley, *Extinction*, 1987)

A disease seems so unlikely, and yet, there was obviously something wrong with them. It's as if they were marked, as if the killer knew who he was after, pointing the finger of death only in certain directions. Some species totally disappeared, others skated through with little apparent harm. It's the puzzle of the ice ages.

The great dire wolf went extinct, but the timber wolf survives to this day. Why the difference? The giant long-horned bison dropped out of the running, but the American buffalo kept multiplying until upwards of 60 million of those wild oxen roamed the American west.

The short-faced bear went extinct. What saved the grizzly? (A carnivore, the short-faced bear had unusually long legs that helped it chase down its prey.)

And what saved so many other ice age animals, such as the musk ox, wolverine, moose, and arctic fox? Santa still has his reindeer, beavers still chew on trees, and the deer and the antelope still play. Who gave them permission to be so special?

Equal opportunity? Forget it. This killer attacked the bigger mammals, the Paul Bunyans of their day, with a vengeance.

And it hated northerners. Animals in the north were almost decimated. It destroyed 39 genera in North America while killing "only" eight genera in Africa and all but ignoring South America and the India and Malay peninsulas. (It's that more-destruction-to-the-north thing again.)

Now, unless you know how many animals can exist in a single genus, 39 genera probably doesn't sound too impressive. Take the genus *Canis* (dogs). It includes several *species*, such as the American timber wolf, coyote, European gray wolf, and domestic dog, which in turn includes several *breeds,* such as poodle, dachshund, pekingese, terrier, spaniel, beagle, and many more. Thirty-nine genera is a whole slew of critters.

There was a North American beaver as big as a black bear. It's now gone. And the armored glyptodont, a bulky ten-foot-long armadillo, is gone too. (Picture a football helmet almost as big as a Volkswagen . . . with legs.)

Talk about big! The ice age king-of-the-jungle made today's lion look like a pussycat. But his size didn't help him one whit. His crown may have been bigger, but he still fell off the throne. And the ice age kangaroo, which stood more than one-story tall (ten feet), also disappeared.

Towering over them all, though, was the giant North American ground sloth. Heavier than a pickup truck (several thousand pounds), it wandered through ancient forests on skateboard feet up to three feet long, and stood as tall as a two-story house. You'd think an animal that big could name its own ticket. No such luck.

Even the scavengers were huge. Ice age vultures cruised through the skies like small airplanes on wingspans almost 12 feet tip-to-tip. With dead mammoths strewn across the land as far the eye could see, they should have been ecstatic . . . except for one minor detail: The vultures were dead—murdered along with the mammoths.

What really has scientists scratching their heads, though, is the timing. Those great "mountains that walk" were killed, not, as intuition would have it, as it was getting colder, but just as the world began turning warm; just as the ice began pulling back.

What killed those ancient behemoths? Some try to blame humans. But there's no way we could have done it. The mammoth died everywhere, even in remote parts of the north that we couldn't possibly have known existed, areas such as Siberia, which covers millions of square miles and is sparsely populated even today.

Besides, we had no reason to hunt many of the animals that died. As far as we know they weren't dangerous. They probably didn't taste very good either.

No, don't blame humans, blame the weather. "I would argue," said anthropologist Dr. Don Grayson of the University of Washington, "that climatic change caused the extinction."

If we blame humans, how can we explain all of the earlier ice age extinctions? The first wave of death rolled across Africa about 60,000 years ago at the peak of a glacial advance when some 40% of the world's larger mammals disappeared. Giant pigs disappeared. Giant baboons disappeared. Giraffes with antlers disappeared. So did three-toed horses. Gone. Totally gone.

And they left in a hurry. They disappeared in "a geological eyeblink," said Windsor Chorlton in the 1983 book *Ice Ages*. Those giants must have had a fatal flaw. But what?

Then, about 23,000 years ago, the blood-thirsty killer returned. And again, big mammals were pushed to the brink of extinction. Some, such as the European forest elephant, fell over the edge, gone forever. (The mammoth was clobbered at that extinction too.)

And that brings us to the extinction of 11,500 years ago. With all the hoopla over dinosaurs, we've lost sight of just how big that extinction really was. "What's the big deal?" we yawn. "So what if a handful of mammoths went extinct?"

It *wasn't* a handful.

Look at the numbers. We've been digging mammoth bones from the ground for thousands of years. Chinese traders began buying Siberian ivory more than 2,000 years ago, said Robert Silverberg in his book *Mammoths, Mastodons and Man*. The Arabs began buying it at least a thousand years ago.

Mammoth bones have been found all over the world, from Alaska to Czechoslovakia to Siberia to Mexico to France. The French found so many bones in one district of Dauphiné province that they called it Le Champs des Géants, "the field of the Giants."

There was so much ivory in some parts of Siberia that it was considered inexhaustible as a coalfield. And it was in prime condition. Some tusks were as clean and white as modern elephant tusks, so white that "they must have come straight out of clean ice."

We still find mammoth bones, even today. An entire cache of mammoth bones was recently discovered on the remote arctic River Berelyakh, a tributary of the Indigirka. Lifted into the site by helicopter, experts from the Yakut Academy of Sciences saw thousands of bones protruding from or lying on the riverbanks, bones so densely packed that they protected the banks from erosion. (John Stewart, *Smithsonian*)

"It's nothing compared to those still buried," said Professor Nikolai Vereshchagin of the Zoological Institute in Leningrad. Heavy erosion on the Arctic coast "washes thousands of tusks and tens of thousand of bones each year into the sea."

It's impossible to know exactly how many mammoths died, but they must have numbered, if not in the millions, at least in the hundreds of thousands. "Some estimates," said Adrian Lister and Paul Bahn in their 1994 book *Mammoths*, "suggest that there are 10 million mammoths still lying in the Siberian deep freeze."

Some Siberian islands are made "almost exclusively" of mammoth bones, said one 19th-century geologist. The Lyakhov Islands seem "almost paved with their bones."

There was so much ivory on Lyakhov's islands (the Lyakhov Islands were discovered by a merchant named Lyakhov) that he mined it; so much ivory that he needed help. He hired so many miners that he ran out of places for them to live and had to build huts for them all. That's a lot of miners.

And a lot of dead mammoths.

Another enterprising guy had a different kind of mine. When Josef Crometschek of Předmostí, in the Czech Republic, stumbled across a vast deposit of mammoth bones on his farm in 1850, he began grinding them into fertilizer. He continued his venture for nearly 30 years before Czech archaeologists finally persuaded him to sell his land to a museum.

The new owners started digging. Six to ten feet down they found a layer of bones nearly three feet thick. After 30 years of indiscriminate mining, it still held the bones, they calculated, of at least a thousand mammoths.

On just one farm.

Crometschek, Lyakhov, and other venturesome individuals like them uncovered mind-boggling amounts of ivory. One ivory hunter—just one, mind you—brought back 20,000 pounds in a single year. But that's nothing. In the town of Yakutsk, the chief ivory marketplace, an average of 50,000 pounds of ivory went on sale each year throughout the 19th century.

Punch that into your calculator! Fifty thousand pounds of ivory a year, every year, for more than a hundred years. That's five million pounds. We're not talking five million pounds of *bones* here, we're talking five million pounds of *ivory!* Five million pounds of mammoth tusks, in just one town, in just one century, and we've been mining that ivory for at least 20 centuries. Entire *herds* of those giant beasts must have been wiped out!

Just as entire herds of dinosaurs died.

Herds of dinosaurs?

Come on now. Dinosaurs didn't herd together like cattle. Did they? Or clump together like a flock of bleating sheep. Did they? Don't destroy all of my old illusions. Weren't dinosaurs supposed to be big, mean, and macho? Weren't dinosaurs supposed to be cold-blooded, lonesome bullies?

Not true. Some dinosaurs were veritable social butterflies, living together in herds so big that they make the American buffalo look like a hermit. An entire herd of dinosaurs was recently discovered by paleontologist Jack Horner. (The inspiration for *Jurassic Park's* fictional hero Alan Grant, Horner is also a professor, and curator of the Museum of the Rockies at Montana State University.)

It was a herd of adult maiasaurs. (Maiasaurs were a type of hadrosaur best known as the duck-billed dinosaur.) Found at Montana's Willow Creek anticline, the discovery was so important that Horner devoted an entire chapter in his 1988 book *Digging Dinosaurs* to "The Herd."[1]

The bones are gray-black in color, said Horner, and strangely battered. Shattered might be a better word. Stretching 1¼ miles east-to-west, and ¼ mile north-to-south, the bone bed contains more than 30 million fragments. There's no way to count the bones, said Horner, but by a conservative estimate it holds the bones of at least 10,000 dead dinosaurs.

No freak accident of nature, said Horner, could have bunched the bones together like that, especially since they aren't in a river or streambed. In fact he doesn't know *what* they're in; some sort of mudstone, he guesses. All massacred at the same time about 80 million years ago, they're buried exactly 18 inches below a layer of bentonite (volcanic ash).

What turned the duckbills into sitting ducks? Maybe they got caught in a mud flow, one worker suggested. But why are the bones in such miserable shape? Some are broken in half, others sheared apart lengthwise. Even more mysteriously, right beside a badly damaged bone will lie a bone that hasn't been touched. How could a mud flow be so selective?

Most perplexing of all, though, are the standing bones. Some bones are standing at attention, sticking straight into the ground. It

---

1. An anticline is a wrinkle, or a fold, in the earth.

looks like a giant's game of mumbletypeg—played with dinosaur bones instead of knives.

Whoa. Floods don't leave bones standing upright, thought Horner. Floods leave bones lying flat, or jumbled together in a mishmash.

Neither do mudslides. What kind of mudslide, he asked, no matter how big, could take a dead animal weighing two to three tons, two to three times more than a modern-day draft horse, and toss it around so hard that "its femur—still embedded in the flesh of its thigh—split lengthwise?"

Why are none of the bones chewed on? Ten thousand dead dinosaurs, doggy paradise, and nothing has gnawed on the bones?

Why are the bones lying east-to-west, the long dimension of the grave? And why are there no babies? Small bones are rare in the main part of the bone pit, said Horner, they're all on the easternmost edge.

Three other digging sites, up to a mile further east, also hold "little" bones from nine-foot dinosaurs. Invariably, said Horner, the bones at the edge are better preserved than those in the middle.

Invariably! Why are the bones at the edge in better shape than those in the center? You'd think any predator with a lick of sense would go for the easy pickings first. Bones of the smaller animals at the edge should be the most damaged, not the least.

The volcanic ash must be the key, thought Horner. Look at Mount St. Helens. "That was a little volcano," he said. "Volcanoes like that were a dime a dozen in the Rockies back in the late Cretaceous." Much bigger volcanoes, he recalled, had erupted south of Willow Creek in the Elk Horn Mountains near Great Falls. Bigger volcanoes had erupted in the Rockies west of the site, too.

A volcanic eruption could explain why no predators had chewed on the bones; the predators had died along with their prey. Then a catastrophic flood moved the bones to their present location and buried them beneath a protective layer of mud.

Maybe. But I don't buy it. Why are small bones found mainly at the edge of the grave? Why are the bones at the edge in better shape than those in the center? Why are the bones black? Why are some bones standing upright? Why are some bones so horribly battered while ones right beside them remain untouched? A volcano wouldn't do any of those things. Neither would a flood.

But a snowstorm could! Especially if it came during a period of massive volcanic eruptions.

Horner's herd was caught in the biggest snowstorm in 14.1 million years . . . the same kind of snowstorm that will soon kill most of us!

With massive volcanic eruptions behind them, and six feet of snow per hour falling on their heads, the desperate maiasaurs stampeded. Eyes rolling, noses snorting, and lungs bellowing, the biggest, the hardiest, the meanest, tore to the front of the pack. (That's how Mother Nature works, isn't it? We call it survival of the fittest. It's really survival of the nastiest. The young ones, the weak ones, the small ones, always get left in the rear.)

But it did no good. Flailing about in frantic attempts to stay above the snow, they instead dug deeper and deeper. Still it kept coming, burying the biggest among them.

It looked as if the meek really would inherit the earth. With no one left to walk on them, the ones in the rear had avoided the onslaught. Now it was their turn to climb the gory ladder of success. God help them, though, if they fell between the rungs. Instant pulverization in the grinding mass below.

Climbing ever higher on the bodies of their fallen comrades, they tried to stay above the ever deepening snow.

Still it kept coming. Four stories deep. Six stories deep. Nine.

All in one day.

Still they kept climbing, nine stories into the sky.

Millions of pounds of live dinosaurs, nine stories deep, biting and scratching and kicking and writhing and jumping on one another in panic would break a lot of bones, I would think, in a lot of funny ways. That's how one bone could get shattered but not the one right beside it.

Reaching the top of the pulsating pile, still panicked, still running, they continued their deadly stampede. What a mistake! They didn't know they were at the edge of a cliff; a cliff built of anguished dinosaur bodies on one side, and nine stories of soft snow on the other, ready to suck them down.

Off the edge of the pile they plowed. And down they went, sinking further and further into the snow. In front of the herd now, but on the bottom, they became living stepping stones for the next wave of their ever smaller brethren.

The smaller they were, the better to crawl on the backs of their heavier kin. The lighter they were, the better to swim through the snow and away from the belching volcanoes.

Exhausted, they stopped to catch their breath and maybe to take a nap. Too bad. Freezing is a peaceful way to go, I hear. You simply drift into a hypothermic sleep . . . and never wake up.

That's why the bones of smaller animals are found at the edge of the pile. That's why the bones at the edge are in better shape than those in the middle. There was no one left to stomp on them.

Then the snow melted. And as it melted, nine stories of dead dinosaurs rotted and fell. Muscles and tendons disintegrated, and bones disarticulated themselves. Tumbling into the snowmelt, they came to rest pointing in the direction of the flow, east-to-west.

Some bones at the top, it seems logical, fell through the nine-story gridwork of rib cages below. Picking up speed as they went, by the time they hit bottom they rammed straight into the mud, just as Jack Horner found them.

What a desolate sound it must have made, with no one alive left to hear it, as the last lonely bone plinked down through the pile, to stand quivering in the macabre mud.

Now we know the dinosaur's fatal flaw. Now we know the mammoth's fatal flaw. They were too heavy. They couldn't climb out of the snow.

Now we know our own fatal flaw!

I see changes in the earth's climate as the most important cause of crises in the history of life.
—STEVEN M. STANLEY

13

· · · · · · · · ·

NINE STORIES
OF SNOW A DAY

· · · · · · · · ·

"A HAIL OF A MESS IN FLORIDA," blared the headline. Two feet of hail fell to the ground in less than half an hour yesterday, the story said, when a freak thunderstorm struck the town of Longwood, Florida. Hail came down so thick and so fast that "it damaged buildings and snarled traffic." (*Seattle Times*, 7 Mar 1992)

Snarled traffic? At two feet deep, that hail stood deeper than the bumpers on their cars. Longwood's traffic wasn't snarled; it was stopped, dead in its tracks.

Ever tried to walk through two feet of snow? Almost impossible, wasn't it? That's why snowshoes were invented. Ever fallen in deep powder? Remember how hard it was to get up? Imagine trying to get up in nine stories of snow. You'd never escape.

Figure it out. Longwood's two feet of hail fell to the ground in less than 30 minutes. That's more than four feet an hour. Let it hail like that for 24 hours, and you'd have 96 feet of hail.

Nine stories!

In one day!

At that rate, even a short six-hour storm would have dumped 24 feet—two and a half stories—of ice on that hapless Florida town. But Florida is mostly a one-story state. A few lonely steeples and tall trees might have poked up through the white blanket of death, but most buildings would have collapsed, buried beneath a barren crust of ice. Who could survive such an onslaught?

To say it can't happen is wishful thinking. Ask the drenched survivors of tropical storm Alberto. When Alberto stalled over Georgia in July 1994, 21 inches of rain fell on parts of the state in one day. If that rain had fallen as snow (just add a zero) it would have measured 210 inches deep—almost 18 feet!

Or look at Houston, Texas. A relentless rainstorm in October 1994 dumped more than 30 inches of rain on the unprepared city in four days. Add a zero, and you'd have 300 inches—25 feet—of snow. Few roofs in the world could handle that kind of weight.

If 30 inches of rain can fall on "The Lone Star State" in four days in October, if a freak thunderstorm can drop two feet of hail on "The Sunshine State" in less than 30 minutes in March, imagine what a *big* storm could do in "Big Sky Country" in the dead of winter.

Granted, Florida and Texas are next to the ocean and Montana isn't. But remember, sea levels were much higher back then. The Willow Creek anticline (where Horner found his 10,000 dinosaurs) was only 100 miles from the coast. Is that so far to ask a vicious winter storm to travel? I don't think so. Neither did the dinosaurs.

Neither did the mammoths. Fully preserved woolly mammoths and woolly rhinoceroses are found even today in relic slabs of ice along the shores of the New Siberian Islands in the Laptev Sea, in the Arctic Ocean.

The fact that mammoths are found encased in ice, and the perfect preservation of their flesh, said the famed 19th-century geologist James D. Dana, "shows that the cold finally became *suddenly* extreme, as of a single winter's night, and knew no relenting afterward." (His emphasis, not mine.)

Whatever killed the mammoths worked fast. It killed them so fast that some are found still standing up, frozen in their tracks. It froze them so fast that their meat, thousands of years later, is fresh enough to eat. Some have undigested food in their bellies.

The first standing mammoth that we know of was spotted in 1839 on the Shangin River, said I. P. Tolmachoff in the journal *American Philosophical Society*.

In 1846, a Russian survey team steaming up the swollen Indigirka River saw yet another. They watched as the mammoth, excavated by the swirling waters, suddenly appeared at the river's edge. "It looked as if the ground thousands of years ago gave way under the weight of the giant," they reported, "and he sank as he stood, on all four feet." A hunter by the name of Boyarski stumbled across another standing mammoth on the Bolshoi Lyakhov Island in 1860.

Those were the first standing mammoths. The first *edible* mammoth (again, that we know of) was found in the side of a cliff near the Beresovka River in northeastern Siberia in August 1900. Its discoverers, hunters from the Lamut Tribe whose dogs had scented the meat, notified the Russian Academy of Sciences, who sent a team to the scene about a year later.

Jutting head-first out of the cliff, with the remainder of its body sealed in a mixture of soil, rock, and ice, the beast was mostly intact, not thawed or rotted away. "The ice surrounding the carcase [*sic*]," said a writer identified only by the initials A. S. W., "was not that of a lake or river, but evidently formed from snow." (*Nature*, 30 Jul 1903)

Building a shack over the giant corpse, then building fires inside the shack for warmth, the Russian scientists unearthed it, all the while keeping careful notes of the beast's condition. Its body, thousands of years after its death, was half standing up. It was "sitting on its haunches in a tumbled-looking way, with its right hind leg thrust forward and its front limbs flexed as if grasping at the ground."

A bull, it had apparently died of suffocation. Its genital organ was erect, a condition only explainable, they said, by asphyxia. (This was not the only ice age animal to suffer such a fate. Blood vessels in a woolly rhinoceros found at the River Vilyui in Siberia, said William Farrand of Lamont-Doherty Earth Observatory, were filled with red, coagulated blood, indicating that it too had died of suffocation.)

Once the body thawed, they dissected it. Inside the gargantuan belly they found more than 30 pounds of undigested food. The food consisted of the same kinds of herbs, flowers, grass and moss that grow in the area today, along with the boughs of tundra shrubs.

And there were seeds in the giant tummy.

The seeds were an important discovery, the Russian scientists said, because they showed that the beast had died in the autumn.[1] When they probed into its mouth, they found its lips and tongue still well preserved, and "between its teeth were fresh flowers." *How's that for dying fast?*

Cutting through its massive rear legs they found dark red meat, fresh-looking meat, as if the animal had died only recently. It looked so fresh that they toyed with the thought of a mammoth-steak cookout, but couldn't quite work up the nerve. Instead, they fed the well-aged meat to their dogs. "The dogs found its taste first-rate." It tasted so fresh and appealing that "they devoured every piece thrown to them." Other reports out of Siberia also tell of dogs eating mammoth meat with no ill effects.

Refreezing the enormous body in the outside air, they sledded it to the Trans-Siberian Railroad and thence to Leningrad, where it was restored and placed on display at the zoological museum.

---

1. Stomach contents of almost all mammoths ever discovered show that they died in the autumn. (Lister and Bahn, in *Mammoths*.)

Another frozen mammoth, found at the Island Mostakh in 1857, was so well preserved, said Tolmachoff, that the natives used its skin, two inches thick, to make dog harnesses, and its fat to lubricate their boots.

Now, if you've ever frozen anything, meat, fish, vegetables— whatever—you know that the quicker you freeze it the better. If you freeze meat slowly, it forms large crystals that burst the cells, thus dehydrating the meat and destroying its flavor. The flesh of many frozen mammoths contains cells that are not burst, indicating that freezing occurred rapidly.

How do you quick-freeze an animal the size of an elephant so that its meat, thousands of years later, is still "dark red and fresh looking?"

With nine stories of snow.

But that's not what the Russian scientists decided. A thin layer of ice must have formed over a deep crevasse, they announced, then got covered by a layer of freshly fallen snow. Stepping on the ice, the unwitting beast broke through and got injured. As it struggled to escape from the pit, it pulled tons of loose snow down on itself. This has been the official explanation for almost 100 years.

Some would have us believe that that's how *all* mammoths died— by falling into crevasses. They'd have us believe that ten million mammoths all became stupid at once, the same not-so-stupid mammoths who had somehow survived for two million years. I hope they were patient, those mammoths; there must have been a waiting line at every crevasse on earth.

Ice age humans may have died just as quickly. Remember Joe Crometshek's mammoth-bone fertilizer mine? Archaeologists didn't quit when they found that three-foot layer of mammoth bones on his farm; they kept digging. At nine feet, they uncovered prehistoric walls built from the shoulder blades and skulls of mammoths. Inside those walls they unearthed the skeletons of about 50 people buried close together in a squatting position.

Must have been a grave, they decided.

A grave? Who buries people in a squatting position? And who builds walls around a grave? Why not a building? Or a stockade? Or a wind-break? If that were the only structure built of mammoth bones ever discovered, I might agree that it was a grave. But it wasn't. At least a dozen sites in the Ukraine contain evidence of mammoth-bone dwellings.

Other such buildings have been found in the Byelorussian Republic. One, now reconstructed, was built of 400 huge bones, tusks, and skulls from 95 different mammoths. The bones formed a dome over which mammoth skins were stretched. More than 70 mammoth-bone dwellings have now been identified.

No, that was no grave. Those people were inside a structure of some sort, hiding from the cold. When you go camping, what do you do? You build a fire. And then? You squat. You squat, stretch out your hands and get as close to the fire as you can. That's what was going on, I believe: 50 lost souls squatting next to a fire. Fifty lost souls huddled together for warmth in a snowstorm that refused to quit.

But forget about people and mammoths if you like, what about the others? The list of animals thawed from the frozen ground is endless. Frozen woolly rhinoceroses (one of which was also standing up), reindeer, giant tigers, giant bison, giant oxen, wolverines, cave lions, beavers, horses, wolves, all have washed from the frigid Siberian soil in countless numbers. Finding frozen carcasses is so common, said Tolmachoff, that "local people usually do not pay any attention."

In some localities, so many bones lie embedded in river banks that the air "reeks with the stench of decomposing bodies." Many more giant bodies lie on plateaus far above existing river levels, indicating that they must been deposited by a gigantic flood. (There's that Noah's Deluge thing again.)

Where is Inspector Clouseau when we need him? Here we find these multi-ton beasts, fast frozen, some standing up, with the killer's icy arms still wrapped around them, in ice "evidently formed from snow" . . . and we can't figure out what killed them? We even know what time of year they died. In the autumn.

What do animals do in the autumn? Many head south for the winter. And that, I believe, is why some fared better than others. With their predators slinking along behind them, many ice age animals had already migrated to warmer climes. But mammoths loved that cold nasty weather. Why leave?

That's why one kind of wolf went extinct and not the other. Those who followed their prey south survived. Those who stayed behind were killed, buried beneath nine stories of snow.

Nine stories of snow a day killed the dinosaurs, nine stories of snow a day killed the mammoths, and nine stories of snow a day will soon kill most of us.

Previous interglacial mild intervals comparable in warmth to the present one lasted only about 10,000 years. The present interglacial has lasted 10,000 years. If it is truly similar to earlier ones—and if man's activities do not alter natural trends—it should be nearly over.

—James D. Hays

14

· · · · · · · · · ·

PACEMAKER
OF
THE ICE AGES

· · · · · · · · · ·

What causes an ice age? Would you believe space dust? The spiral arms of our galaxy contain dust, said W. H. McCrea of the University of Sussex, U.K. As our solar system hurtles through the arms, the dust hides the sun and temperatures plummet. No, say skeptics, the sun would burn the dust as fuel and shine even brighter. The earth would get hotter, not colder.

Another theory, a good one, came from Maurice Ewing and William Donn of Lamont-Doherty Earth Observatory. It's cold enough in the Arctic right now, they said, to cause an ice age. All we need is more moisture.

Right now! But where would the moisture come from?

Every now and then, they theorized, the Arctic Ocean must rid itself of ice, thereby allowing warm North Atlantic waters to mix with the colder water. Thus heated, evaporation increases, albeit only slightly, more moisture rises into the sky—and *voila!*—more snow.

They were so darn close.

Deglaciation begins, they proposed, when temperatures drop and the Arctic Ocean freezes. With the moisture supply cut off, the ice melts and sea levels rise again. Subsequent testing showed, however, that the Arctic Ocean hasn't been ice-free for millions of years.

And that brings us to the astronomical theory.

Major glaciations correlate with our galactic orbit, said William Forbes of Cornell University in 1931. (A full revolution, called a cosmic year, takes about 282 million years.) Umbgrove broached the idea again in 1947.

How could our galactic orbit cause an ice age? Biologic, tectonic, and geomagnetic forces, said Johann Steiner and E. Grillmair of the University of Alberta, could all be affected by the gravitational changes caused by the slight eccentricity of the orbit.

Our galactic orbit is not a perfect circle, Steiner explained, it's elliptical. Our distance from the galactic center therefore varies from about 32,600 light-years when we're closest, to about 38,000 light-years when we're furthest away. We're approaching our closest position right now, and will reach it within eight million years (plus or minus four million). Since previous major glaciations occurred about every 140 million years, "a causal relation appears plausible."

Peaks in radiogenic strontium also correlate with our galactic orbit, said Steiner. So do peaks in lead. (Uranium decays into lead.) At least two Precambrian ice ages correlate with two of the six lead isotope events in prehistoric Canada. All lead isotope events, Steiner theorized, will eventually be tied to glaciation. (*Precambrian Research*, 1978)

G. E. Williams, editor of *Megacycles: Long-Term Episodicity in Earth and Planetary History*, agreed. On average, said Williams, major ice ages have occurred twice per cosmic year (but not at the end points of the elongated part of the ellipse).

Ice ages not only correlate with our galactic orbit, they also correlate with our celestial orbit. The seeds for that discovery were planted about 450 years ago when Nicolaus Copernicus, a Polish cleric, proved that the earth is not the center of the universe. The earth's only job, said Copernicus, is to orbit the sun in a circle.

No, our orbit is not a circle, said the astronomer Johannes Kepler, it's a stretched-out oval. (So sometimes we're closer to the sun than others.)

The Paris mathematician Joseph Adhémar refined the theory even further. The sun isn't where you'd expect it to be in that oval, said Adhémar; it's off-center, which means we spend a few more days each year away from the sun than close to it. Those two orbital variations, Adhémar believed, together with the earth's tilt in relation to the solar plane, cause ice ages.

Ice ages seesaw back and forth, north to south, said Adhémar, depending on where we are in the orbit. When the North Pole tilts away from the sun, it gets cold in the north. When it tilts toward the sun, it gets cold in the south. And since the sun isn't in the center of the orbit, the North Pole tilts away from the sun longer than it tilts toward it, making it even colder.

So far so good. But then he blew it.

The Antarctic ice sheet has such a strong gravitational attraction, said Adhémar, that it sucks water out of northern seas making southern waters bulge. Then they freeze. When the water warms up, the base of the ice melts while the top looms over the sea like an overgrown mushroom teetering on a fragile stem of ice. When the stem melts, the mushroom collapses, and huge tidal waves rush north at hundreds of miles per hour. No proof has ever been found for this part of his theory.

Along came Urbain Leverrier with yet another piece of the puzzle. Our orbit's shape constantly changes, said the Parisian astronomer. Starting as an almost perfect circle, it slowly stretches into an oval. Then, due to the gravitational pull of the planets, it collapses back to nearly a perfect circle again.

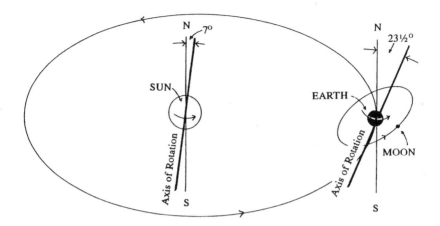

The sun's axis of rotation tilts 7° away from perpendicular to the earth's orbit, while the earth's axis of rotation tilts 23½°. The sun is off center of the earth's orbit around it. The sun and the earth rotate counterclockwise, the earth orbits counterclockwise, and so does the moon. (After Robert E. Noyes, in *The Sun, Our Star*.)

Called orbital eccentricity, or orbital stretch, it takes about 100,000 years for the orbit to stretch from a circle to an ellipse and back to a circle again. Today our orbit is only slightly eccentric, about one percent, but it varies from a low near zero to about six percent. We're about 11 million miles further from the sun when the stretch is greatest.

In 1864, the astronomical baton was passed to James Croll, a self-educated Scotsman. Like Adhémar before him, Croll knew that the earth's axis of rotation is not perpendicular to the solar plane. Today we're tilted at 23½°, but the tilt slowly increases to about 24½°, then decreases to about 22°. The complete shift, back and forth, takes about 41,000 years.

Croll also knew that our axis of rotation wobbles like a top, tracing a clockwise circle around true north. Called axial precession, it takes about 25,800 years to make the full circle. Precession occurs, say scientists, because the sun and moon exert a gravitational pull on the earth's equatorial bulge. Rotating objects such as tops and gyroscopes also precess. So does Mars.

To understand this phenomenon, picture the globe spinning around a long stick (the axis of rotation). Tilted away from true north, the top of the stick traces a circle around the North Pole, while the bottom makes an identical trip around Antarctica.

As our axis of rotation moves, it constantly points toward a different star. If the stick were longer, it would paint an imaginary circle on the heavens. The process of painting that circle on the celestial ceiling is called precession of the equinoxes.

This is not a new theory. The Greek philosopher/scientist Hipparchus first discovered precession of the equinoxes in the second century B.C. when he compared his own astronomical measurements to those made by Timocharis some 150 years earlier. But he mistakenly concluded that the stars were moving in a circle through the sky.

Eighteen centuries later, Sir Isaac Newton solved the rest of the riddle. Our orbit around the sun also revolves, said Newton. The orbit itself revolves backwards, or counterclockwise. Precession of the equinoxes, the time it takes to paint that imaginary circle on the heavens, therefore takes about 23,000 years. It's like waiting for someone on a merry-go-round. You'll reach them sooner if you walk toward them.

This is deadly important. The equinoctial precession cycle is a killer . . . and it's about to strike again!

Today when viewed from the northern hemisphere, the stars seem to rotate around Polaris at the end of the handle of the Little Dipper. That's why it's called the Pole Star; because the North Pole points toward it. But in 2,000 B.C. the North Pole pointed toward a spot halfway between the Little Dipper and the Big Dipper. In 4,000 B.C. it pointed toward the end of the handle of the Big Dipper. Twelve thousand years from now it will point toward a different star, toward Vega, and in 23,000 years it will point toward Polaris again.

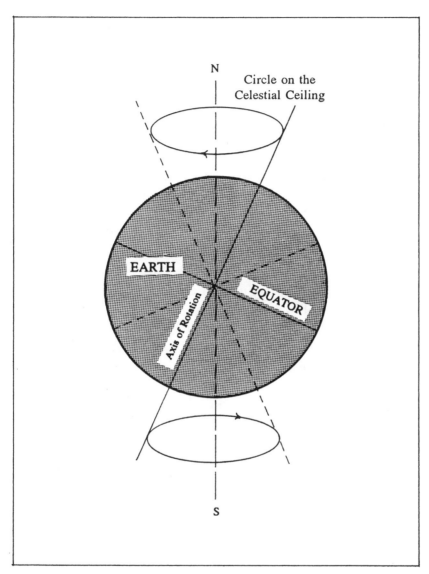

Precession of the Equinoxes.

If the North Pole tilted away from the sun while we were far away, said Croll, an ice age would occur in the north and a warming trend in the south. During the other half of the cycle, an ice age would occur in the south.

Less sunlight in winter would make more snow, and more snow would reflect more sunlight back into space, making it even colder. Croll called this self-perpetuated cooling trend "positive feedback." We call it the albedo effect. It was the same conclusion that Adhémar had reached earlier, but for very different reasons.

Icy periods in our history, called glacial epochs, were separated by periods of warmth (interglacial epochs). The last glacial epoch began about 250,000 years ago, said Croll, and ended about 80,000 years ago. We've been in an interglacial ever since.

Brilliant. Croll was a hero. Scientists had recently discovered that glacial drift (the rocks and debris dropped by glaciers) had not all been deposited at the same time, but consisted of many distinct layers, often separated by layers of peat.

In those layers of peat they had found seeds and leaves of plants that could not possibly have survived an ice age. The layers seemed to prove Croll's theory. They certainly proved that our world had endured many ice ages, not just one.

Croll wouldn't remain a hero for long, though.

Honeymoons are supposed to begin at Niagara Falls, not end there. But when geologist Sir Charles Lyell calculated the rate of erosion at the Falls, Croll's honeymoon with science was over. The ice had begun retreating about 30,000 years ago, said Lyell, not even close to Croll's figure. Then the state geologist of Minnesota, Newton Winchell, calculated how fast the Falls of St. Anthony on the Mississippi River near Minneapolis were receding. The last ice age, said Winchell, ended a mere 8,000 years ago.

The divorce was final. By the time Croll died in 1890, his theory was close to extinction.

But it would make one more comeback under the inspired guidance of Milutin Milankovitch, a young Yugoslavian engineer. Milankovitch acquired his doctorate in engineering at Vienna's Institute of Technology, then worked five years as a practical engineer building concrete dams and bridges. He enjoyed the work, but had a nagging feeling that he should be doing something "more important."

In 1911, during a night of tippling with a friend (they downed several bottles of wine, he later admitted), his mind suddenly became expansive and bold. He decided to master the entire universe. He would develop a mathematical theory to describe the climates on Earth, Mars, and Venus, at different latitudes, at any time.

A tall order. No one had ever calculated the distribution of sunlight over the wobbling, tilting planets. Never mind that he had made his momentous decision with a mind perhaps a tad fuzzy, he pursued his new-found goal with single-minded devotion.

Working on his own, and dusting off calculations previously made by others, Milankovitch decided, as had Croll before him, that three orbital properties determine where and how long the sun shines on our planet: equinoctial precession, orbital stretch, and the three-degree change in axial tilt.

In theory, each sunshine minimum should cause an ice age. (This was all done mathematically; he left it to geologists to find the proof.) The minima formed an uneven pattern, said Milankovitch. There should have been ice ages 25,000, 72,000 and 115,000 years ago. His theory matched fact amazingly well. Or so people thought.

Now Milankovitch was the hero.

But those layers of peat that had once helped prove the theory would later come back to haunt it. When radiocarbon dating first appeared in 1951, geologists joined the dating game with abandon. They dated so many things that a new magazine, *Radiocarbon*, was spawned just to keep track.

"Let's date a layer of peat," somebody said. Uh-oh, something was dreadfully wrong. A layer of peat from Farmdale, Illinois, dated at 25,000 years old. What was going on? Milankovitch had called for an ice age 25,000 years ago—but peat only forms when it's warm.

Must be a quirk, let's look somewhere else. But they found 25,000-year-old peat everywhere they looked. They found it in other areas of the Midwest, they found it in eastern Canada, they even found it in Europe. There was ice when it should have been warm, and heat when there should have been ice. Climatic changes conflicted almost every step of the way with the theory. In theory, 72,000 years ago should have been a time of minimum sunshine and maximum ice.

But in practice? Seventy-two-thousand years ago the ice had ended in Quebec, painfully far north of its previous southern extent. It had *not* been a time of maximum ice. Milankovitch was wrong! Bummer. By 1969, the astronomical theory was on the ropes once again.

But it kept one foot in the ring. Arguments against it were based on incomplete records taken on land. What about the oceans? Dropping hollow pipes into the seafloor, scientists found fossils of microscopic floating organisms called foraminifera, or forams.[1]

The top layer, today's layer, held warm-water forams as expected. The second layer, which had formed *abruptly* about 11,000 years ago, held mainly cold-water forams, but one kind of foram was conspicuously absent. The third layer held cold-water forams again, but in different proportions.

The missing foram, *Globorotalia menardii*, meant the second layer had formed when the water was too cold for menardii to survive. The top layer, rich with menardii, had formed after the ice had melted. The third layer, which also contained menardii, had formed during the warm period just before the ice age began. The layers contained evidence of at least nine ice ages (which was good news for the theory; Milankovitch had predicted nine sunshine minima).

Oxygen isotope data supported the findings. Pioneering work by Dr. Cesare Emiliani at the University of Miami, later refined by Nicholas Shackleton, a Cambridge geophysicist, and John Imbrie, a professor of oceanography at Brown University, showed that seawater contains two types of oxygen atom isotopes. One isotope, oxygen-18, is heavier than the other, oxygen-16.

As seawater evaporates, molecules containing the lighter oxygen-16 tend to evaporate faster, giving them more opportunity to fall to the earth as snow. Enrichment of oxygen-18 molecules in ocean sedi-

---

1. Forams (plankton) still live in the seas today. They range from about one millimeter in diameter to sometimes as big as a pinhead. Chalk and nummulitic limestone are composed chiefly of their calcareous shells.

ments is therefore a strong indicator of an ice age. (It doesn't necessarily mean that temperatures were colder, mind you, just that total ice volume had increased.) The amount of salt in the water also affects menardii, scientists found. Saltier seas meant that more freshwater had been locked up on land as ice, another indicator of an ice age. (*Ice Ages: Solving the Mystery*, John and Katherine Imbrie)

While oceanographers studied the seas, geologist Dr. George Kukla analyzed wind-blown deposits of silt called loess (pronounced "lus" as in "lust") in a brick quarry in Czechoslovakia. During each ice age, Central Europe had become a dry and barren polar desert, with nothing to stop the cruel winds that roared across the land. Those winds deposited the layers of loess.

Each time the ice had melted, the climate grew warmer and wetter than today. Broadleaf trees appeared and grew into far-flung forests. As the leaves fell they formed rich layers of fertile soil. When the ice returned, the forests pulled back, swinging back and forth across Central Europe like clockwork.

From fertile soils to loess, then back to fertile soils, each swing of the pendulum lasted about 100,000 years. Kukla found evidence of at least 10 soil-loess cycles, vindicating Milankovitch once again. It was the same cycle that oceanographers had found in the Caribbean.

Inexplicably, each cycle had ended abruptly. The switch from dry, dusty and freezing, to moist leafy forests, occurred so abruptly that the change showed up as distinct lines, called marklines, in the quarry walls.

Now we had evidence from both land and sea. Major ice ages, no question, had occurred about 100,000 years apart. But what caused the cycle? Since orbital stretch marches to the same beat, it must be the culprit. Let's see if axial-tilt and equinoctial precession also show up in the record.

In 1970, James D. Hays of Lamont-Doherty Earth Observatory spearheaded a project called CLIMAP (Climate: Long-range Investigation Mapping and Prediction) to map the history of the oceans. Frequencies higher than the 100,000-year cycle were clearly visible. Axial tilt showed up, and so did equinoctial precession, with a major cycle of 23,000 years.

Equinoctial precession: Pacemaker of the Ice Ages, they called it.

"It is concluded," said CLIMAP participants Hays, Imbrie, and Shackleton, "that changes in the earth's orbital geometry are the fundamental cause of the succession of Quaternary ice ages." "The long-term trend over the next several thousand years is toward extensive Northern Hemisphere glaciation." (*Science*, 10 Dec 1976)

The astronomical theory of the ice ages was right!

This time, it appeared, the issue was settled for good.

But is the issue ever settled for good? Sure, all of those astronomical properties influence the weather. But three huge problems still remain.

The first, is that ice ages do not seesaw back and forth, north to south, the way they're "supposed" to. The last ice age, said T. V. Lowell in *Science,* "ended suddenly and simultaneously in both polar hemispheres."

"Indisputable evidence exists," said Wallace Broecker of Lamont-Doherty Earth Observatory, "for strong warming in both hemispheres at the end of the last ice age." "The coincidence is not easily explained." (*Milankovitch and Climate,* Part 2)

It makes no difference where you look, said Broecker. Hawaii, East Africa, New Guinea, Columbia, Chile, Tasmania, the Alps, New Zealand, the Cascades, the Himalayas; "on all mountains, regardless of the geographic setting . . . the snow line descended by about one kilometer [six-tenths of a mile]." "The timing is a puzzle." (*Scientific American*, Jan 1990)

The second problem is the magnitude of the change. So much ice accumulated, say scientists, or so much disappeared, that it can't be explained by the small change in the amount of insolation (the amount of sunlight reaching the earth's surface). There must have been "an amplifying mechanism which is not yet understood." (Peteet, Rind and Kukla, 1992)

But the third problem is the biggest problem.

Each of the things we've been talking about here—axial tilt, equinoctial precession, and orbital stretch—occur gradually, over several thousands of years.

So why do ice ages begin and end *abruptly?*

Our present interglacial has probably nearly run its course. While the world worries about the prospect of a nuclear winter, we may suddenly find the seas lowering by 100 meters or more as the next glacial stage advances over Europe, Asia, and North America.

—S. Warren Carey

15

. . . .

NOAH'S DELUGE

. . . .

Are we sure that ice ages end abruptly? Incredibly sure. The last ice age ended so fast, said Windsor Chorlton in his 1983 book *Ice Ages,* that it caused one of the greatest floods in history. "In some areas, said Chorlton, "coastal forests were submerged so quickly that instead of decomposing, the doomed trees were preserved, virtually pickled in the briny water." Sea levels rose 350 to 400 feet.

Glacial melt-water poured into the Gulf of Mexico so fast about 11,600 years ago, said marine biologist Cesare Emiliani, that sea levels surged, causing widespread flooding of low-lying areas. Since the date matches Plato's date for the sinking of Atlantis, said Emiliani, it may explain the deluge stories so common to many Australasian, American, and Eurasian cultures.

Indeed, carbon dating shows a short, rapid, major climate change around 11,000 years ago, said Wallace Broecker of Lamont-Doherty Earth Observatory. Ocean temperatures increased 6° to 10°C (10° to 18°F), and sea levels rose. So did lake levels in normally dry regions of the southwestern United States. (*Ice Ages: Solving the Mystery*, John and Katherine Imbrie)

At the same time in the northeast, said Broecker, torrential floodwaters from Canada's glacial Lake Agassiz raced across the Great Lakes area, then rushed down the St. Lawrence River to the North Atlantic. South of the St. Lawrence, in the lower Connecticut River Valley, floodwaters rose 150 to 200 feet above the present high-water mark. (G. F. Wright, *The Ice Age in North America*)

Further west, great melt-water torrents and flash floods tore across western Canada, eroding the soil so deeply that they exposed vast dinosaur boneyards from the end-Cretaceous.

Thus were born Alberta's badlands. At one time a series of deltas and river flood-plains leading to the ancient inland sea, land along Alberta's Red Deer River is now a barren moonscape of steeply eroded channels, gullies, and hoodoos stretching from Red Deer City to the Saskatchewan border.

Meanwhile, great walls of water raged across the northwestern United States. A massive flood cut channels hundreds of feet deep into Columbia Plateau basalts at the end of the last ice age, said geologist J Harlen Bretz of the University of Chicago. (That's not a typo; there's no period after the "J" in J Harlen.)[1]

Blasting out of the Clark Fork River valley of western Montana and racing across northern Idaho at 10 cubic miles an hour, the

---

1. How many millions of humans, I wonder, washed out to sea during those monster floods? How many tiny towns along the banks of every river and stream disappeared, along with all vestiges of their very existence? And if sea levels rose by 350 to 400 feet, and coastlines extended 100 miles further out, how many cities sank into the sea to be pickled along with the trees?

waters reached depths of 800 feet (about 80 stories) as they plunged through the Wallula Gap on the Oregon-Washington line, then tore down the Columbia on an unstoppable rampage to the Pacific. Major scabland rivers and the Columbia were hundreds of feet deep, and reached widths of 50 miles. (The flood left thousands of *scabs* of exposed basaltic rock in its wake, hence the name *scablands*.)

Sweeping away as much as 100 to 200 feet of topsoil in many locations, the flood denuded fully 2,000 square miles of the Columbia Plateau of its silt cover and loess, leaving only steep-walled, trench-like valleys (coulees) up to 400 feet deep as barren reminders of its awesome power.

Water poured into the Snake River so fast that an enormous back-rush tore eastward *up* the river, rehandling gravel in the bottom of Snake Canyon and redepositing it upstream as far away as Lewiston, Idaho, a distance of 70 miles. Further east, it dumped sand and silt up to three stories deep in the towns of Juliatta and Arrow in the lower Clearwater Valley.

"The flood arrived catastrophically," said Bretz. "It was a great wall of water, with its crested front constantly outrunning and over-running its basal portion." Up to 1,300 feet high, it poured over the tops of nearby ridges as giant waterfalls and cascades up to nine miles wide, and rolled boulders many feet in diameter for miles. In the Willamette Valley, flood-rafted boulders have been found at 450 feet. In the uplands above the Umatilla basin, they've been spotted more than 1,000 feet above present-day water levels.

The flood ended as fast as it began, in a matter of days. It left giant river bars now standing as mid-channel hills more than 100 feet high, and deposited a 200-square-mile gravel delta at the junction of the Willamette and Columbia River valleys. Portland, Oregon, and Vancouver, Washington, now sit atop a portion of that delta.

But that's not the end of the story. The great thickness of some valley-floor accumulations, said Bretz, show that such floods are a recurring phenomenon. "At least 8 successive catastrophic outbursts correspond with the 9 glacial retreats that took place in northern Montana, Idaho, and eastern Washington."

One such flood swept down the Columbia River some 80,000 to 100,000 years ago. Another, about 33,000 years ago, corresponded with the break-out of Lake Bonneville. (A precursor to Utah's Great Salt Lake, and almost as big as Lake Michigan, Lake Bonneville stood 1,100 feet deeper than today. It covered most of Utah and parts of Nevada, Wyoming, and Idaho. If Lake Bonneville should recharge itself—*a process that I believe has already begun*—Salt Lake City would lie under a fifth of a mile of water.)[1]

When Bretz first presented his views in 1923, people laughed. But he mounted "a stout, lonely, and dogmatic defense" (his words, not mine), until today his theory prevails. "These ideas have stood the test of detailed ground-level mapping," he would later boast, "the test of aerial photography, the test of inspection by critics, even scrutiny by the satellite *Nimbus I.*"

"Virtually all geologists now believe," said Stephen Jay Gould in his 1980 book *Panda's Thumb*, "that catastrophic floods cut the channeled scablands."

But where did the water come from? It came from a vast lake in the north, scientists decided, bigger than all five Great Lakes put together. They named it glacial Lake Missoula.

Up to 1,400 feet deep, glacial Lake Missoula was contained behind an ice dam of unprecedented proportions. The ice dam, which we call the Purcell Lobe, was created by a 45-mile-wide tongue of ice reaching south out of Canada to block the Clark Fork River near the present town of Sandpoint, Idaho. When the ice dam collapsed, a great wall of water roared across the western plateau at several hundreds of miles per hour.

---

1. Great Salt Lake recently reached an historic high, threatening to swamp Salt Lake City International Airport, two major railroad tracks, two interstate highways (I-15 and I-80), many homes and businesses, and has inundated Saltair resort. Its rise has been temporarily slowed by a $60 million project to pump water from the lake. (*Salt Lake Tribune,* 30 Oct 1994)

The Great Lakes also stand at record heights. Fueled by unusual amounts of precipitation, water levels in the Great Lakes have risen 5 to 7.5 feet (1.5 to 2.0 meters) since 1964 (Chernicoff).

Similar catastrophic emptying (jökulhlaup) of ice-dammed lakes occurred, at about the same time, around the world—from the Scottish Highlands, where earthquake activity accompanied the outburst of a glacier-dammed lake in Glen Roy (Ringrose)—to southern Scandinavia, where the Baltic Ice Lake poured into the North Sea—to the sudden draining of the Pur and Mensi ice-dammed lakes in western Siberia. (Dawson, 1992)

But why did those ice dams collapse? And why did so many other huge floods occur?

Why did lakes levels rise so precipitously in Mexico's Lake Chalco and Lake Chiconahuapan? Why did lake levels rise so precipitously in other intertropical lakes such as Ethiopia's Lake Ziway-Shala or in Ghana's Lake Bosumtri? (Street-Perrott and Perrott)

By the same token, why did East Africa's Lake Victoria and Lake Albert (Lake Mobutu Sese Seko) suddenly overflow, causing severe flooding on the Nile? (Dawson)

Because huge rainstorms, I submit, caused by the warming evaporating seas, breached the dams and caused the floods, both in the north and in the south.

Anecdotal evidence supports the idea of warmer seas. In Mexico, the sacred book *Popol-Vuh* speaks of ancient boiling seas. So do the Mexican *Manuscript Calchiquel* and the *Manuscript Troano*. So does the Iranian *Zend-Avesta*. So does British Columbian Indian folklore. (From *Worlds in Collision*, by Immanuel Velikovsky.)

Tales of warming seas also come from the Makah Indians, who live at Neah Bay on the northwest corner of Washington's Olympic Peninsula. They tell of a time when "the sea rose without any waves until it submerged Cape Flattery." "The water on its rise became very warm, and as it came up to the houses, those who had canoes put their effects in them, and floated off with the current." (From *Strange Planet, a Sourcebook of Unusual Geological Facts*, by William R. Corliss, p. E2-50.)

The rising seas penetrated deep into the continents, said the Swiss scientists Edith Kristan-Tollmann and Alexander Tollmann, in their 1993 book *Und die Sinflut gab es doch*. "In many places the flood is said to have invaded the continents at boiling heat."

And that's why I'm worried . . . because our seas are warming right now!

Ocean temperatures are rising, not only in Monterey Bay (see Chapter 10), they're rising in other locations, too. Water temperatures at Point Conception north of San Diego, said Dean Roemmich and John McGowan of the Scripps Institution of Oceanography, have risen 1½°C (2° to 3°F) since 1951.

As water temperatures rose, sea dwellers died. Zooplankton populations in the area declined by 80%. Since zooplankton is the main diet of many species of fish, including anchovy, jack mackerel, sardines, and others, the Scripps researchers worry that such "stunning" losses could eventually affect the entire food chain. Birds, too, have been seriously hurt, said McGowan. "We don't know if this warming of the ocean is man caused or part of a natural cycle." (*Burlington* [Vermont] *Free Press*, 3 Mar 1995)

Declining salmon runs in the Pacific Northwest may also be caused by warming seas, said oceanographer Bill Pearcy of Oregon State University. "Without cold water, plankton and other creatures that make up the lowest link on the food chain fare poorly, leaving salmon little to eat." (From "El Niño the Weathermaker," Douglas Gantenbein, *Popular Science*, May 1995)

Now take a wild guess as to what's happening in our skies in conjunction with our warming seas.

Water-vapor levels are rising!

Using balloon-borne water-vapor detectors, Samuel Oltmans and David Hoffman of NOAA's Climate Monitoring and Diagnostics Lab in Boulder, Colorado, tracked water-vapor concentrations in the stratosphere from 1981 to 1994. Water-vapor levels are rising rapidly, they found, "about one percent per year." (*Nature*, 9 Mar 1995)

No wonder we're seeing ever larger floods!

They credit the rising water-vapor levels to increased methane levels. Stratospheric chemical conversion, they believe, changes each methane molecule into two molecules of water vapor. Since humans supposedly caused the rising methane levels, by implication then, humans caused the rising water-vapor levels.

But the implication is wrong.

We didn't cause the rising methane levels (as you'll see in Chapter 16), we didn't cause the rising water-vapor levels (they're caused by warming, evaporating seas), and we're not causing the warming seas . . . they're caused by underwater volcanism.

An immense submarine lava flow was recently discovered off the coast of South America near the East Pacific Rise, said Stanley Chernicoff in his 1995 book *Geology*. The flow, believed to have been extruded into the sea within the last 25 years, "contains approximately 15 cubic kilometers of basalt, enough to pave over the entire U.S. interstate-highway system to a depth of 10 meters, or 35 feet."

Many lava flows along the East Pacific Rise are so fresh, said the French scientist Jean-Marie Auzende, that "hot shimmering waters still rise from the ocean floor." This means, said Auzende, that they erupted within the last decade—some within the last few months. (*Eos*, 20 Dec 1994)

Hawaii's underwater volcano Loihi may also be headed for a major eruption. Loihi (pronounced low-EE-hee) is located 20 miles southeast of the southern tip of the island of Hawaii. A seamount, Loihi rises nearly two miles above the seafloor. Even then, its peak remains 3,200 feet below the surface of the ocean. During a single week in July 1996 Loihi generated more than 1,500 earthquakes. Scientists assumed that a massive undersea eruption could soon occur . . . or already had. (*Seattle Post-Intelligencer*, 26 Jul 1996)[1]

And we wonder why our seas are warming?

We've got it backwards. Warming seas *cause* climatic change, they're not the result.

You can make the case, agrees oceanographer Nicholas E. Graham of Scripps Institution of Oceanography, that sea temperatures do affect the climate. The recent warming trend in global temperatures has been driven almost entirely by increasing tropical sea temperatures, said Graham. Warmer ocean temperatures "cause increased evaporation and increased precipitation, thereby heating the atmosphere through the release of latent heat." (*Science*, 3 Feb 1995)

---

1. Unbelievably, although Loihi is the most active volcano in the world, its activity remains unmonitored as of this writing.

The amount of heat transferred from seas to air is "enormous," said Wallace Broecker of Lamont-Doherty Earth Observatory.

The top 10 feet (three meters) of the ocean contain as much heat as the entire atmosphere, said Kevin Trenberth of the National Center for Atmospheric Research in Boulder, Colorado. "It makes sense, therefore, that the atmosphere will follow the ocean's lead." (From "Tropical Trouble: Two decades of Pacific warmth have fired up the globe," by Richard Monastersky, *Science News*, 11 Mar 1995)

Arun Kumar and his colleagues at the National Meteorological Center in Camp Springs, Maryland, reached a similar conclusion. Comparing air temperatures to sea surface temperatures from 1950 to 1993, they found that, almost without exception, air temperatures rose and fell in perfect phase with El Niño. "This suggests," they said, ". . . that warm oceanic conditions in the tropics contributed to the observed global warming trend during this period." (*Science*, 28 Oct 1994)

It's not global warming, it's *ocean* warming!

But forget warming seas and global warming for a moment; let's get back to J Harlen Bretz and his 80-story floods. Whether those floods came from bursting ice dams, warming seas, or both, Bretz missed an important part of the story: He thought the floods had recurred eight times.

I think they recurred millions of times.

Consider. What happens during a flood? Erosion; then deposition (i.e. sedimentation). Such a process of flood-caused erosion and sedimentation can be seen in the stratigraphic record, not eight times, not hundreds of times, but literally millions of times through history.

And each of those millions of floods occurred in phase with equinoctial precession.

Wilmot Bradley noticed the link between sedimentation and precession in 1930 when he discovered an approximate 22,000-year cycle in the Green River formations of Utah, Colorado, and Wyoming. But he refused to believe his own eyes. It would be easy to say the changes are due to precession, said Bradley, except for one thing: "The change from one kind of rock to the other is abrupt."

Roger Anderson of the University of New Mexico reported a similar cycle in 250-million-year-old sediments of the Delaware Basin in southeast New Mexico.

Triassic rock formations in northern Italy tell the same story, said Lawrence Hardie of Johns Hopkins University (with colleagues Bosellini and Goldhammer). It's a story of repeated floods in sync with equinoctial precession. There is "firm evidence," said Hardie, for a regular pulse in sea level oscillations about every 10,000 years.

The link between sedimentation and equinoctial precession is now "firmly established," said F. J. Hilgen, and is underlain by similar climatic oscillations. (*Earth Planet. Sci. Lett.,* 1991)

Indeed, the sedimentation cycle appears to have been controlled by precession for the last 2.2 billion years, said J. P. Grotzinger of Lamont-Doherty Earth Observatory. *"Either alpine or very limited continental glaciation may have been responsible."* (Emphasis mine.)

Volcanism tracks the precession cycle, too, and has done so for millions of years. End-Cretaceous clays, said Robert Rocchia of the French Atomic Energy Commission, contain iridium peaks caused by volcanoes about 10,000 years apart. The Columbia River basalts also erupted in 10,000-year pulses, said Tony Irving at the University of Washington. So did Germany's West Eifel volcanic field (Schnepp and Hradetzky).

And that brings us back to the seas.

If volcanism tracks the precession cycle, and if 80% of all volcanism occurs underwater (as NOAA says), does submarine volcanism track the precession cycle too?

It appears to. New underwater volcanoes erupt from the seafloor about every 10,000 years, said Allan Cox and Robert Brian Hart in their 1986 book *Plate Tectonics, How it Works.*

Now, what would happen to our weather, do you suppose, if thousands of underwater volcanoes should erupt in all of the world's oceans at once, pumping untold amounts of red-hot basalt into the seas?

I think we're about to find out.

*Orbital forcing might be the cause of long-term
secular variation in the geomagnetic field.*
—TOSHITSUGU YAMAZAKI & NOBORU IOKA

16

. . . .

THE NEXT ICE AGE—NOW!

. . . .

Ice ages not only end abruptly, they begin abruptly too. The proof lies buried in a peat bog in Alsace, northeastern France. Called the Grande Pile peat bog, it contains pollen from the last glacial, the last interglacial, and the Holocene (the last 10,000 years).

Pollen deposited in the peat bog during the last 300 years of the Eemian Period came from trees that grow in warm weather. (The

Eemian Period was a time of warmth similar to today. It separated two ice ages, lasted about 11,000 years, and ended about 115,000 years ago.) (Kukla)

When the Eemian Period ended, it ended suddenly, said Geneviève Woillard of the Université Catholique de Louvain. Vegetation in the peat bog changed "radically" and warm-weather trees completely disappeared. The change occurred "in less than 20 years."

Every ice age began fast.

"When we looked at the records of past temperate intervals," said Woillard, "we found abrupt shifts in forest composition at the end of all previous interglacials." *All!*

The change from one kind of forest to the other occurred so rapidly, said H. H. Lamb, in *Climate, history and the modern world*, that the time elapsed was less than the error margin for radiocarbon dating. The current elm disease in Europe and North America, Lamb noted, may be part of a similar pattern. The next ice age "could quite well be imminent."

The earth underwent "climatic havoc" at the end-Eemian, said geologists Paul J. Hearty and A. Conrad Neumann. The transition from greenhouse to icehouse was a "madhouse." Sea levels surged 20 feet above modern-day levels, then plunged at least 50 feet in less than a century. Today's rapidly rising sea levels, they worry, may be an indication that we're "in a madhouse again." (From "High Tidings: Ancient, erratic changes in sea level suggest a coming swell," by Christina Stock, *Scientific American,* Aug 1995)

The two geologists ascribe the sudden rise in sea level to a glacial surge. But I think vast amounts of red-hot basalt poured into the seas, forcing them upward and heating them. Excess moisture then rose into the sky, more precipitation fell to the ground—and presto!—instant ice age—the same thing that will soon happen to us.

More proof that ice ages begin abruptly comes from the European Greenland Ice-Core Project (GRIP). Drilling cores almost two miles deep into the ice in central Greenland, GRIP members obtained a detailed climate record for the past 250,000 years. The end-Eemian ice age began "catastrophically," they found. Worldwide temperatures plummeted 20°F almost overnight.

"It was as though the climate of Nome, Alaska, suddenly descended on San Francisco," said astonished science reporters. (Bellevue *Journal American*, 15 Jul 1993)

Isotopic changes were violent, said Willi Dansgaard of the University of Copenhagen. A similar almost instantaneous change from a climate warmer than today to one of full-blown glacial severity, Dansgaard added, occurred about 90,000 years ago.

It made no difference what test they ran, *every* ice age began with a bang. Electrical conductivity tests showed that some transitions had occurred as quickly as three years.

Strontium ratios in the seas changed abruptly, too. The changes show no lag time, said Philip Froelich in *Nature*, they're in phase with the growth and decay of the ice. But that's "impossible," said Froelich, "strontium has a residence time in the sea of several million years, too long to explain such rapid rates of change."

We've long assumed that strontium came from erosion, that rivers washed it from the land and into the seas. (Himalayan river flows, for example, are rich in strontium-87.) But the drastic changes can only be explained, said Froelich, if erosion almost stopped during the ice age, which is "an unlikely occurrence." "We're left with an uncomfortable, Sherlock Holmes-like choice between the impossible and the improbable."

The strontium cycle is far more dynamic than we thought, agreed Steven Clemens, John Farrell, and Peter Gromet of Brown University. Strontium ratios change so rapidly that there may be glacial-age strontium sources not yet accounted for. (You bet there are strontium sources not yet accounted for! It rains from the sky during magnetic reversals.)

Dramatic spikes in other elements appear in the record, too. Radioactive carbon-14 levels increased 300% to 400% at the end of the last ice age, said the French scientist Alain Mazaud. This "strongly" reinforces the hypothesis, said Mazaud, that geomagnetic variations are the major source of long-term variations in the abundance of carbon-14.

Another drastic increase in carbon-14 (400% to 500%) occurred 18,000 to 22,000 years ago, said Edouard Bard of Lamont-Doherty

Earth Observatory. The most reasonable explanation, said Bard, "involves magnetic modulation."

Carbon dioxide levels also increased (about 30%) at the end of the last ice age (Neftel *et al.*). And it wasn't the only time. The carbon dioxide (CO_2) record, said J. M. Barnola, exhibits a cyclic change corresponding to precession. It's called the carbon cycle.

Same with beryllium-10. Twice the normal amount of radiogenic beryllium-10 is found in 60,000-year-old ice, said G. M. Raisbeck of the Laboratoire René Bernas in Orsay, France. Another spike occurred about 35,000 years ago. "The dramatic difference," said Raisbeck, "may have been caused by changes in magnetic intensity, and may be related to the Lake Mungo magnetic excursion." (More about the Lake Mungo excursion later.)

Another spike in beryllium-10 occurred at the Brunhes magnetic reversal. Other spikes occurred at 105,000, 90,000, 68,000, and 23,000 years ago. The most recent spike, two to three times normal, occurred about 11,000 years ago at the end of the last ice age (Beer, 1984). (With so much debris falling from the sky, the earth *must* be expanding. But that's another book.)

Methane levels also rose (they doubled) at the end of the last ice age. Again, it wasn't the only time. Changes in methane levels are cyclic, and can be linked to orbital variations (Chappellaz). Peaks in sulfate, nitrate, and chloride also occurred at the end of the last ice age (Herron and Langway).

Same with "soot." While exploring ancient caves for his doctoral thesis, John S. Kopper of Columbia University discovered a layer of black carbon deposited near the end of the last ice age. It must be soot, Kopper theorized. Maybe early humans managed their crops by burning.

Sure, blame it on humans, that's who we tried to pin the mammoth extinction on, too. Maybe it *is* soot. Maybe it *was* caused by humans. (I'm guessing it was created in the sky, and that, too, is another book.) But should we take the rap for all of the other debris swirling through those ice age skies?

No way. Nor should we take the rap for today's increases in the same elements. Carbon dioxide levels, along with methane, hydrocarbons, sulfur and nitrogen oxides, and others, are increasing daily, said Dixy Lee Ray (with Lou Guzzo) in their 1990 book *Trashing the Planet*. The rate of increase is "substantial," about one percent per year.

Again, many scientists blame humans. "But it's not as simple as that," said Ray, former chairman of the Atomic Energy Commission. "Such increases have occurred in the past without any help from us at all, and this time is probably no different. Most likely, the causes were and still are colossal cosmic forces quite outside human ability to control."

Which brings us back to magnetic reversals and equinoctial precession. Put them together, and you've found that colossal cosmic force that Ray and Guzzo were talking about.

Scientists agree that equinoctial precession affects our magnetic field. They just don't know why. And many agree that magnetic field intensity waxes and wanes in a cycle. "Dipole intensity fluctuations are periodic," said Allan Cox, in *Plate Tectonics and Geomagnetic Reversals*, "with a period of about 10,000 years."

Ominously, we're nearing the end of just such a period right now.

During the last 2,000 years, said Peter J. Smith in *The Earth*, dipole field strength has fallen 50%. (Walter Elsasser of the University of Utah pegged the decline at 65%.) Five percent of the decrease occurred during the last 100 years alone. "This suggests," said geophysicists McFadden, Merrill, and McElhinny, ". . . that we are experiencing either a fluctuation from the mean behavior, *or a precursor to a new reversal attempt*." (Emphasis mine.)[1]

When will the reversal occur? "If present trends were to continue," said Harwood and Malin, of the Institute of Geological Sciences in Sussex, U.K., "the field would reverse in about 2230 A.D."

It could reverse tomorrow!

Here's why. As our axis of rotation makes its 23,000-year circle around the North Pole, magnetic intensity slowly increases. Then it decreases, every 11,500 years. Up and down it goes, turning in and

---

1. McFadden and McElhinny are with the Australian Geological Survey. Merrill is at the University of Washington.

out of sync with the solar system's magnetic field like a giant rheostat switch in the sky, the same way that a light bulb glows dimmer or brighter as you turn a dimmer switch.

Normally, magnetic intensity rises and falls gradually. Sometimes it rises or falls within the 11,500-year cycle itself. But toward the end of each cycle, it drops through the floor.

And that's when the trouble begins. In a study of lava flows at Steens Mountain, south central Oregon (which erupted *during* a reversal, by the way), Michel Prévot, Edward Mankinen, Robert Coe, and C. Sherman Grommé found that magnetic intensity had fallen to less than 10% of today's in less than one year.

Perhaps in less than two months!

During a follow-up study in 1989, Coe and Prévot found that the field had reversed at the rate of three degrees per day.

Perhaps in only three weeks!

Not content with their earlier findings, Coe and his colleagues took another look. The earth's magnetic field had reversed at "the astonishingly rapid rate," their new study found, of six to eight degrees per day.[1] Not only did it reverse, it fluctuated. Rapid fluctuations occurred many times during the reversal, said Coe. "Enhanced external [magnetic] field activity . . . from the Sun might somehow cause the jumps." (Coe, Prévot, and Camps, 1995)

Such oscillations have been noted in other locations, too. Valet, Laj, and Langereis found rapid magnetic field fluctuations during a reversal in western Crete, while Michael D. Fuller and his colleagues at the University of California, Santa Barbara, found rapid fluctuations in the Tatoosh Intrusion on Washington's Mount Rainier. (From "Ancient Magnetic Reversals: Clues to the Geodynamo," by Kenneth A. Hoffman, *Scientific American*, May 1988)

Those magnetic field fluctuations, I submit, generate massive surges of electricity in and above the earth. Electrotelluric forces go

---

1. The speed of the reversal is calculated by comparing magnetic field direction in different parts of the same lava flow. By estimating how long it took the center to cool as opposed to the faster-cooling edges, a time frame can be established.

crazy, thousands of cubic kilometers of basalt pour into the seas, radioactivity falls from the sky, and carbon dioxide levels shoot through the roof.[1] How long the process takes, and whether it becomes a full or aborted reversal (an excursion), depends on where we are in our stretched-out orbit around the sun, which in turn determines how strongly its magnetic lines of force affect us and therefore how destructive it is.

Then snap! It's that torque thing again. Since the earth can't turn over, our magnetic field reverses. Or, depending on where we are in our orbit, it continues fluctuating, generating ever more electricity until we move out of alignment again. It happens twice per rotation: once on "this" side of the precessionary wobble, once again on the other, every 11,500 years. It's the same thing that happens to our solar system every 141 million years in its orbit around the galactic center, but on a much smaller scale.

It's a celestial game of orbital tag . . . and we're it.

The alignments alternate: one in the fall or winter, and then, 11,500 years later, one in the spring or summer. Underwater volcanoes heat the seas, and thousands of cubic kilometers of moisture rise into the sky. If it's warm at the time, huge rainstorms—Noah's Deluge-type rainstorms—melt the ice and the world floods. If it's cold, the rain turns to snow, and the self-perpetuating albedo effect begins. That's why ice ages begin or end abruptly every 11,500 years. And that's why the next ice age could begin any day.

Today's decreasing magnetic field strength could also explain our recent rise in volcanic activity. "We are now living in a period of vastly increased volcanism, the greatest in the past 500 years," said Dixy Lee Ray (with Lou Guzzo) in their 1993 book *Environmental Overkill, Whatever Happened to Common Sense?*

"The 1980s were one of the most volatile decades of volcanic devastation ever recorded," said David Mazie of the *National Geogra-*

---

1. Indeed, there is a direct link between carbon dioxide levels and warming seas. "Sea-floor hydrothermal activity accounts for about 29 percent of the total ocean contribution to atmospheric CO_2." "[It's] a significant factor in the present-day CO_2 budget." (Owen and Rea, 1995)

phic News Service. New Guinea, Columbia, Mexico, the Philippines, the United States, Zaire, New Zealand, Samoa, the Cape Verde Islands, Guatemala, Russia's far east, all have seen increased volcanic activity in the past few years.

And it appears to be getting worse.

Many "extinct" volcanoes are stirring to life as you read this. During the last 50 years, said Robert L. Christiansen of the USGS, the Yellowstone caldera in Yellowstone National Park has uplifted at rates nearly comparable to some active volcanoes and may be headed toward another major ash-flow eruption.

Or look at Mount St. Helens. Earthquake activity at Mount St. Helens increased from less than 10 events per month in January 1995 to about 100 per month in September 1995, say geophysicists at the University of Washington. They worry that additional small explosions could occur without warning. (*Washington Geology*, Dec 1995)

Or consider southern Oregon, where Mount Mazama once soared into the sky. When it erupted about 6,700 years ago, more than a mile of mountain disappeared, leaving a giant water-filled caldera that we call Crater Lake. "Recent dives to the lake bottom," said Stanley Chernicoff, "reveal that hot water and steam are being vented continuously, suggesting that Mount Mazama may be entering a new eruptive sequence."

Look, too, at Iceland, where Mount Helgafjell on the island of Heimaey recently erupted, its first eruption in 5,000 years. Or look at the 1994 Rabaul eruption in New Guinea. A spate of earthquakes just prior to the eruption gave seismologists their main clue that something was about to happen. Immediately sounding the alarm, they saved more than 30,000 lives. Now considered a seismological triumph, Rabaul is also deemed a "caution" to the U.S. West Coast from California to Alaska.

Why a caution?

Because 12 years ago, earthquake activity at Rabaul increased dramatically, then slowed. When activity resumed, it resumed with a vengeance, giving only one day's warning of its impending eruption. California's Mammoth Lakes area is similar to Rabaul in that it too endured an alarming series of earthquakes about 12 years ago, then slowed. As at Rabaul, scientists fear that heightened activity could return at any time. Rabaul is a "sobering lesson for us," said Bill Ellsworth of the USGS.

Sobering indeed. *If the 1980s were one of the worst decades of volcanic devastation ever recorded . . . and if 80% of all volcanic activity occurs underwater, imagine what's happening in our seas!*

Massive floods in the making! That's why ocean temperatures are rising. That's why sea levels are rising. How many miles of basalt must pour into our seas, and how far must our magnetic field decline, before we acknowledge that we have a problem on our hands?

We're certain that volcanic activity increases during times of low magnetic intensity. Take Germany's West Eifel volcanic field. A "remarkable proportion" of its eruptive activity, said Schnepp and Hradetzky, occurred when magnetic field strength was only 15% to 60% of today's. (*Jour. Geophys. Res.*, 1994)

Deep sea cores tell the same story. The cores "show clearly," said oceanographers Kennett and Watkins, that volcanic maxima occurred when geomagnetic polarity changes were taking place.

Major volcanism occurred at the Jaramillo magnetic reversal, and again at the Brunhes reversal of 780,000 years ago. Germany's West Eifel volcanic field, with its 240 different eruptive centers, saw its main activity *during* the Brunhes reversal (Böhnel). Massive volcanism occurred *during* the Brunhes reversal in New Zealand (Kennett). And a three-mile-tall monster volcano south of present day Mono Lake, California, disappeared in one of the largest eruptions to ever occur on the North American continent. (A seething magma chamber still lies beneath the valley floor.)

Is it just a coincidence that *Homo erectus* (Java Man, or Peking man) appeared at the same reversal? Or that several new marine species appeared at the same reversal (Opdyke)? Or that Tasmania's Darwin Crater formed at the same reversal (Grieve and Robertson)? I don't think so.

Major volcanism also occurred at the Delta magnetic reversal of 630,000 years ago (Champion), when Yellowstone pumped out more than 1,000 cubic kilometers of ash in a matter of hours or days

(Christiansen). The Big Lost magnetic reversal of 510,000 years ago (Champion) also saw major volcanism.

Decreasing magnetic field strength could also explain our recent increase in seismic activity. India, Chile, Indonesia, Bolivia, Turkey, New Guinea, Japan, Greece, Mexico, Burma, Sumatra, China, Ecuador, Columbia, the Philippines, parts of Texas, parts of the U.S. East Coast, and much of the U.S. West Coast from Alaska to California, have seen escalating earthquake activity.

Indeed, California may be the harbinger of things to come. "The pace of earthquakes throughout the Los Angeles region is increasing ominously," said seismologist Kate Hutton. Southern California has been struck by 24 temblors of magnitude 5 or greater since 1932. Eight of them since 1987. "The reason for the increase in activity is a mystery." (*Seattle Post-Intelligencer*, 18 Jan 1994)

Earthquakes and rising land; two tectonic processes in phase with precession of the equinoxes.

For proof, look at Barbados, the easternmost island in the Lesser Antilles in the Caribbean. Barbados is terraced. From the air, the terraces look like a huge flight of stairs.

Two theories exist as to what formed the terraces. Each theory begins with the same premise, that the island periodically rose from the sea during a major earthquake. Each time it rose, the first theory goes, one reef died and a new one grew at a lower point on the island. The second theory holds that each terrace was carved from a single large fossil reef. Each time the island rose, wave action sculpted a new terrace.

In 1965, Professor Robley K. Matthews of Brown University had the terraces dated. The lowest one dated at 82,000 years old, the second one dated at 103,000, and the top one dated at 122,000. The steps, said Matthews, were sculpted in sync with precession of the equinoxes. (Broecker *et al.*, 1968)

Other parts of the world also rose in sync with precession. Each terrace on New Guinea's Huon Peninsula, for example, formed when it abruptly rose above sea level during a major earthquake in sync with precession (Bloom, Broecker, Chappell, Matthews, and Meso-

204 NOT BY FIRE BUT BY ICE

lella). The Florida Keys also rose in sync with precession (Broecker and Thurber). So did the Bahamas and the Ryukyu Islands. Other rapidly rising reefs have been dated at 66,000 ± 4,000 years, others at 48,000 years—all in sync with equinoctial precession. Even the exposed beds in southern Scandinavia south of Göteburg, which uplifted about 10,000 years ago, rose in sync with precession.

And the pace is picking up.

"The rates of modern movements," said Officer and Drake, "are significantly greater than the average rates over the past 130 my." These movements may be episodic, said Officer, "with a cycle as short as 10,000 years." (*Tectonics*, 1985)

What would make huge chunks of land halfway around the world from each other rise in phase with equinoctial precession?

Electromagnetic forces—unleashed by magnetic reversals.

What magnetic reversals?

Didn't our most recent reversal occur at the Brunhes/Matuyama boundary? No, not even close. At least ten reversals or excursions have occurred since then.

In 1967, Norbert Bonhommet and J. Babkine discovered a reversal in lava flows at Laschamp and Olby, at Chaîne des Puys (chain of volcanoes), in central France. Our magnetic field reversed about 20,000 to 30,000 years ago, they said, then remained reversed for about 10,000 years. They called it the Laschamp reversal. Is it just a coincidence, Bonhommet asked, that the return to normal corresponded with the end of an ice age?

Though later research placed the Laschamp event at about 47,000 years ago, its discovery made us aware that other reversals might have occurred.

And they have . . . many times.

The most recent reversal, the Gothenburg excursion, occurred about 12,350 years ago (Mörner and Lanser).[1] During the excursion, magnetic intensity dropped dramatically, and magnetic inclina-

---

1. Merrill and McElhinny placed the Gothenburg at 8,000 to 14,000 years ago (in *The Earth's Magnetic Field*). Kenneth Creer placed it at 11,000 to 14,000 years ago. (*Earth Planet. Sci. Lett.*, 1976)

tion moved 180°. It also fluctuated, making wild swings of up to 80° (Kopper).

Another excursion, the Mono Lake excursion, occurred about 23,000 years ago (Kukla, Berger, Lotti, and Brown). During the Mono Lake event, magnetic intensity fell ten times faster than normal (Liddicoat and Coe).

Before that came the Lake Mungo excursion of about 33,500 years ago (Barbetti and Flude), and prior to that was the "real" Laschamp event of about 47,000 years ago, when magnetic intensity fell to less than 15% of today's. (All reversals and excursions show major decreases in intensity. Roperch, Bonhommet and Levi, 1988)

See the cycle? Those excursions struck just like clockwork every 11,500 years.

And they've been doing it for millions of years. Reversals about 10,000 years apart, said volcanologist Vincent Courtillot, have been found in the 65-million-year-old Deccan Traps. Indeed, 10,000-year hiatuses between lavas of opposite polarities are observed "frequently." (Watkins)

However, they're often disregarded. Magnetic intensity fluctuations of from two to 30,000 years duration appear in the marine record as "tiny wiggles," said Steven Cande and Dennis Kent of Lamont-Doherty Earth Observatory, and are therefore easy to overlook. We believe that this type of behavior (of tiny wiggles), they said, "may have characterized the geomagnetic dynamo throughout the Cenozoic [the last 65 million years]." (Cande and Kent, 1992)

I think we'll eventually find millions of such "tiny wiggles" in the geologic record.

But now, here's the topper: *Catastrophic cooling and rapid ice build-up accompanied many of those excursions.*

The Gothenburg excursion coincided with a period of short-term ice and snow, said Michael R. Rampino of NASA. So did the Lake Mungo excursion, when rapid cooling immediately followed a period of warmth.

The Mono Lake excursion coincided with a period of glaciation, said Rampino, as did the Blake magnetic reversal at the end-Eemian.

So did Biwa I, a reversal about 195,000 years ago. Biwa II, a reversal about 286,000 years ago, also shows glaciation, as does Biwa III (about 390,000 years ago), when rapid ice build-up followed a period of warmth similar to today's.

Each of those catastrophic cooling episodes, said Rampino, "may have been triggered by a magnetic excursion." "The Earth's magnetic field may be directly modulated by precession." (*Geology*, Dec 1979)

And there you have it.

Polarity reversals, equinoctial precession, and ice ages, all march to the same drummer. As do extinctions, new species appearance, volcanism, and rising land. Toss in the specter of massive floods, 30-story tsunami (tsunami is both singular and plural), and radioactivity falling on your head, and you've got the picture.

Look at the number of catastrophes that have befallen our planet in almost perfect sync with equinoctial precession during the last 127,000 years (127 kya) alone:

127 kya - Barbados and other islands rise. Ice age ends abruptly. A period of warmth similar to today's begins (W. Broecker). *Homo sapiens neanderthalis* suddenly appears (Hoyle).

115 kya - Blake magnetic reversal. Carbon-14 and strontium peaks. Ice age begins abruptly following a period of warmth similar to today's.

103 kya - Ice age ends. Barbados, other islands rise during a major earthquake. Beryllium spike.

91 kya - Ice age begins catastrophically after a period of warmth similar to today's. Beryllium spike.

80 kya - Ice age ends. Barbados, other islands rise during a major earthquake. Lake Missoula flood.

69 kya - Yellowstone lava flow (Christiansen). Ice age begins abruptly. Rising reefs and terraces (Bloom). Beryllium spike.

58 kya - Mass extinction (40%). Giant pigs, giant baboons, three-toed horses . . . all gone. Beryllium spike. Rising reefs. Major volcanism (Dawson). Ice age ends.

47 kya - Laschamps magnetic excursion. Ice age begins abruptly. *Homo sapiens* suddenly appears (Royal Tyrrell Museum). Rising reefs. Arizona's Meteor Crater forms. (A deep source magnetic minimum, perhaps "unrelated to a meteor" underlies Meteor Crater.) (Regan and Hinze)

34 kya - Lake Mungo (Australia) magnetic excursion. Extinction. Short-term ice build-up, then ice age ends abruptly. Lake Missoula flood. Lake Bonneville flood. Beryllium spike. Intensive volcanism. Neanderthal disappears (Eldredge).

23 kya - Mono Lake magnetic excursion. Extinction. Ice age begins abruptly. European elephant disappears. Mammoths clobbered. Carbon-14 spike. Beryllium spike. Major volcanism. The Mono Lake event actually straddles a layer of ash (Liddicoat).

11 kya - Gothenburg magnetic excursion. Mass extinction; 72% of large mammal species go extinct, whereas only 10% of small mammal species disappear. Rapid and severe ice build-up, then ice age ends in less than 20 years (Dansgaard). Today's warm period begins. Spikes in carbon-14, beryllium, strontium, and many other elements. Exposed beds south of Göteburg uplift. Earthquakes in Scotland, Canada, and Scandinavia (Ringrose). Tectonic uplift along Nile, possibly worldwide (Dawson). Ocean temperatures rise 10° to 18°F. Lake Bonneville flood. Alberta's badlands flood. Gulf of Mexico flood. Connecticut River flood. Lake Missoula flood. St. Lawrence River flood. Nile River flood (Fairbridge). Many other superfloods. Worldwide volcanism. Mexican volcanism "firmly" dated at 11,580 ± 70 years (Street-Perrott). Most recent eruption of Germany's West Eifel fields (Lamb). Laacher See eruption in East Eifel fields (Bogaard). Mount St. Helens ash interlayered with Lake Missoula flood deposits, indicating simultaneous events (Chernicoff). Glacier Peak, Washington, eruption (Dawson). In Alaska and Siberia, ash lies interspersed through piles of mammoth bones themselves

(Hibben). Alaskan volcanism of tremendous proportions" (Hibben). Mt. Katla, Iceland, eruption (Dawson).

0 kya - And that brings us to today, a period of rising sea levels, rising lake levels, rising sea temperatures, rising seismic activity, rising volcanic activity, and the worst floods in more than 500 years.

Germany, Belgium, Japan, Costa Rica, Nepal, Norway, Brazil, India, France, China, Italy, Spain, the Philippines, North Korea, Turkey, Luxembourg, Morocco, Vietnam, the Netherlands, Laos, Honduras, Thailand, Ghana, Bangladesh, Pakistan, the United States, Canada, all have seen gigantic floods in the past few years, gigantic floods caused by ever larger storms.

Annual precipitation has increased steadily in Europe and the Soviet Union since the mid-19th century. In the United States, precipitation has increased "markedly" in the last 30 years alone. (R. S. Bradley, *Science*, 1987)

In the years since Bradley wrote those words, precipitation has increased even more. California, Washington, and much of the midwest have been inundated by record-breaking floods. Seventeen inches of rain fell on Slidell, Louisiana, during a May 1995 storm. One month later, three inches of rain *per hour* fell on parts of Virginia. And in December 1995, parts of the coast near Monterey, California, received 22 inches of rain in 12 hours. If that rain had fallen as snow (just add a zero) Monterey would have been buried under 220 inches of snow—18 feet—in 12 hours.

Meanwhile, eight tropical storms and eleven hurricanes, the most during any year since 1933, pummelled the Caribbean. Why? Blame our warming seas. "The warmer the ocean gets," said A. E. (Sandy) MacDonald of NOAA, "the meaner the tropical storm." (In "This is Global Warming?" by Sharon Begley, *Newsweek*, 22 Jan 1996)

Or consider the hail. In early 1995, hail the size of baseballs strafed northeastern Oregon, hail the size of softballs pounded parts of Texas, and basketball-size hail (one hailstone reportedly weighed 33 pounds) hammered parts of China's Guandong province. (*Seattle Times*, 21 Apr 1995)

These storms are not accidental, folks. It's only a matter of time until one of them strikes in the winter. And then?

Instantaneous ice age.

I think it has already begun.

Here we sit, worried about global warming, while, in 1995, the earliest snowstorms in memory walloped Sweden and Denmark, stranding people in their cars and homes. Here we sit, worried about global warming, while, in 1995, the earliest avalanches ever reported swept down the flanks of the Himalayas.

Here we sit, worried about global warming, while, in 1995, record early snowstorms roared through North Dakota, Montana, Nevada, Colorado, Wyoming, and Kansas, and radio stations played Bing Crosby's "White Christmas" in mid-September.

Here we sit, worried about global warming, while, in early 1996, "The Blizzard of '96" smothered the U.S. East Coast under a killer blanket of snow (the most on record), and so much snow fell on northern Japan that the city of Sapporo asked for military assistance.

Here we sit, worried about global warming, while, in July 1996, freak snowstorms fell on parts of South Africa and France, other parts of France flooded, the worst floods in its history inundated Chicago's southern suburbs, and huge floods tore down the Saguenay River in northern Quebec.

Here we sit, worried about global warming, while, in July 1996, massive floods destroyed more than 8,000 homes in North Korea, wiped out more than 100,000 homes in China, and drove the mighty Yangtze to its third highest level ever recorded.

Here we sit, worried about global warming, unaware that our glaciers are growing.

Our glaciers are growing!

It's one of the best kept secrets in the world today.

Field observations over the past 20 years show that the Antarctic ice sheet is growing, said Charles Bentley, director of the University of Wisconsin's Polar Research Center. Greenland's ice sheet is also growing, said Bentley.

Growing! Not melting!

Satellite measurements confirm Bentley's findings. Measurement of ice sheet elevations by satellite altimetry show that the Greenland ice sheet is growing thicker in both the ablation (melting) and accumulation zones. "Present accumulation rates [about eight inches per year]," said H. J. Zwally of NASA, "are larger than the long-term average."

Think about that! More than five million square miles of Antarctica and 700,000 square miles of Greenland—an area twice the size of the continental United States—are covered by ice. Adding eight inches of ice per year to such a vast area is a major accomplishment.

Canada's glaciers are growing, too. Some glaciers on Canada's Baffin Island are as large or larger than at any time during the past 33,000 years; perhaps the past 60,000 years, said Gifford H. Miller of the University of Colorado.

Same around the world.

Glaciers in coastal mountain ranges such as in Alaska and Norway, said L. R. Mayo and R. S. March of the USGS, show evidence of recent growth. (*Annals of Glaciology*, 1990)

Why, oh why, are we ignoring these warnings?

Here we sit, waiting for the weather. Here we sit, knowing that ice ages have begun or ended abruptly, every 11,500 years, for the past half-million years at least.

Here we sit, with a decreasing magnetic field, increasing volcanism (both underwater and above), increasing earthquake activity, rising land, rising CO^2 levels, rising methane levels, rising sea and lake levels, rising sea temperatures, declining zooplankton and fish populations, more moisture in our skies, growing ice sheets, bigger floods, and harsher blizzards every day.

And we're worried about global warming.

During the past two million years, no interglacial lasted more than 12,000 years. Most lasted only 10,000. Statistically speaking then, the present interglacial is already on its last legs, tottering along at the advanced age of 10,000 years.

—JOHN AND KATHERINE IMBRIE

17

. . . . .

THE END

. . . . .

Repent! Repent! Repent! Sometimes I feel like one of those stringy-haired guys in a robe wandering the streets of Manhattan with a sandwich board and an attitude. "Repent! Repent! Repent!" he yells. "The world will end tomorrow!"

Yeah, sure, another doomsday prophet asking for your faith.

But I'm not asking for your faith, I'm asking for your mind. I want you to understand, with every ounce of rationality in your body, that we've been worshiping the wrong god.

Just as so many ancient cultures offered needless sacrifices to their gods in vain attempts to control the weather, so have we been offering our own needless sacrifices to The Great God Global Warming. Methane levels are rising? Throw another industry to the lions.

Now, I'm not denying that methane levels in our skies, along with carbon dioxide, hydrocarbons, and sulfur and nitrogen oxides, are rising, or that ozone levels are dropping. Nor am I denying that we contributed to the changes. But our contributions are minor. Do we honestly believe that we can change the weather, when we don't even know if it will rain tomorrow?

If we're causing the problem, why did similar changes occur at the end of the last ice age? Why did similar changes occur about 23,000 years ago? Or 35,000 years ago? Or 58,000 years ago? Or 115,000 years ago? Or 780,000 years ago? Or 65 million years ago?

If we're causing today's rising methane levels, who's causing the high methane levels on Jupiter, Neptune, Saturn, Uranus, and Pluto? Who's causing the recent increases in earthly seismic activity? Or in earthly volcanic activity?

By the same token, if we're causing today's rising ocean temperatures, why did ocean temperatures rise so dramatically at the end of the last ice age? Or at the dinosaur extinction of 65 million years ago, long before humans set foot on this planet?

Because humans had nothing to do with it.

Re-read one of those horror stories about how we greedy humans are causing global warming. But this time look for the waffle that's almost always buried near the end of the article. "Oh, by the way," it will whisper, "these changes may be caused by a natural cycle."

You bet they're caused by a natural cycle! Dixy Lee Ray knew it all along.

Let me hammer it home once again. "Such increases have occurred in the past without any help from us at all," said Ray, "and this time is probably no different. Most likely the causes were and still are colossal cosmic forces quite outside human ability to control."

Put equinoctial precession and magnetic reversals together, and you've found that colossal cosmic force that Ray was talking about. It's the same force that causes extinctions and creates new genetic combinations. It's the same force that drives islands and mountains ever higher into the sky, the same force that pumps millions of cubic kilometers of fiery basalt into our seas, and the same force that's pumping catastrophic amounts of basalt into our seas right now.

It's not global warming, it's *ocean* warming. We didn't cause it, and we can't change it. All we can do is survive it.

Remember, it's cold enough in the Arctic right now to cause an ice age, said Maurice Ewing of Lamont-Doherty Earth Observatory. All we need is more moisture.

Well . . . now we have it. It's coming from our warming evaporating seas. That's why the next ice age could begin any day.

And it's all caused by magnetic reversals.

That there is a link between magnetic reversals and ice ages is undeniable. Look at the record. Ice played a major role at almost every extinction in history. Climatic cooling was the "dominant agent" at the Cambrian extinction, said Steven M. Stanley of Johns Hopkins University, as it was at the Ordovician, the Permian, the Devonian, the Carboniferous, the mid-Miocene, and yet again at the end-Miocene.

The end-Triassic, end-Jurassic, end-Silurian, and end-Eocene were times of glaciation, and four extinctions during the Cambrian can also be attributed to glaciation. The Precambrian extinction saw extensive glaciation, the dinosaurs died during a short period of glaciation, and so did the mammoths.

And what did those glaciations have in common?

Magnetic reversals.

The end-Carboniferous, end-Ordovician, end-Permian, end-Cambrian, and end-Triassic were times of frequent magnetic reversals or changes in polar wander, while the Cretaceous, Eocene, and Miocene actually ended with magnetic reversals.

At least twelve magnetic reversals can be linked to glaciation during the last three million years alone. A reversal about three million years ago marked the onset of glaciation, said James D. Hays of Lamont-Doherty Earth Observatory. A reversal about two million years ago marked the onset of glaciation, as did a reversal about one million years ago. The Jaramillo reversal marked the onset of glaciation, and so did the Brunhes reversal.

The Biwa I, Biwa II, Biwa III, and Blake reversals coincided with glaciation, said Michael R. Rampino of NASA, as did the Lake Mungo, Mono Lake, and Gothenburg excursions. Each of those catastrophic cooling episodes, said Rampino, "may have been triggered by a magnetic excursion."

Magnetic reversals correlate, we now know, not only with glaciation, but also with volcanism.

Cores collected by the USNS *Eltanin* "show clearly," said oceanographers Kennett and Watkins, that peaks in volcanism occurred *during* magnetic reversals. In 14 samples from eight different cores, only one did not show volcanism at a reversal. Mauritius, Rodriguez, and the Réunion Islands all had volcanic activity at reversals. So did Nunivak Island.

So did New Zealand. Of five distinct layers of ash on New Zealand, one lies near the end-Jaramillo reversal, while two actually straddle the Brunhes/Matuyama boundary. "We find these observed changes," said Kennett, "difficult to accept as purely coincidental."

Look, too, at Steens Mountain in south-central Oregon, which erupted during a reversal. Or look at Germany's West Eifel volcanic field, which saw its main activity during the Brunhes/Matuyama reversal. Or consider the Yellowstone eruption of 630,000 years ago, which also occurred at a magnetic reversal.

And that brings us to the crux of the matter.

If volcanism correlates with magnetic reversals, and if 80% of all volcanism occurs underwater (as NOAA says), underwater volcanism must go bonkers during magnetic reversals, pumping vast amounts of red-hot basalt—2,150°F hot—into the seas.

Underwater volcanism. That's why sea levels rose so abruptly at the end-Eemian (and why they're about to do so again). That's what heated the waters that flooded Cape Flattery (in the Makah Indian legend). That's what heated the K-T seas and evaporated them. That's what heated the mammoth's seas and evaporated them. That's where the moisture comes from to cause glaciation. That's why ice ages correlate with magnetic reversals.

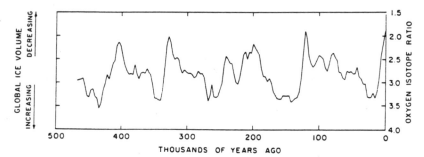

Pacemaker of the ice ages. Changes in global ice volume during the last 500,000 years, as determined from CLIMAP isotopic measurements. (Chart is from John and Katherine Imbrie's book *Ice Ages, Solving the Mystery,* by permission of Enslow Publishers. Data from J. D. Hays *et al.,* 1976, by permission J. D. Hays.)

And that's why the next ice age could begin any day.

But now, for those who still refuse to believe, I've saved one last salvo. Earlier, we discussed a project called CLIMAP (Climate: Long-range Investigation Mapping and Prediction), formed to map the history of the oceans. CLIMAP participants James D. Hays, John Imbrie, and Nicholas J. Shackleton tracked changes in global ice volume for the last 500,000 years. Then they drew a chart (above) of their findings. Take time to study their chart and you'll see that ice ages do begin or end, just like clockwork, every 11,500 years.

Now take a second look at the chart. See the sharp peaks every 100,000 years or so? Each peak marks the abrupt end of a period of warmth similar to today's and the catastrophic beginning of a new ice age.

See where we are today? (At the far right side of the chart?) We're at the tip of the highest peak ever, teetering on the knife-edge of disaster. We haven't been that high on the chart for half a million years. And do you see what happened—without exception—every time we got that high on the chart?

Instantaneous ice age!

Leave out magnetic reversals if you like. Leave out equinoctial precession. Leave out the recent dramatic increases in seismic and volcanic activity. Leave out all of those things. Look only at the glaciation cycle and you'll see that we still have a major problem.

Regardless of what caused it, the fact remains that the last ice age ended almost exactly 11,500 years ago. And the fact remains that the next ice age could—no, *should*—begin any day.

If we keep ignoring the glaciation cycle, we'll go to our deaths—as have so many millions of hapless animals before us—never knowing what hit us.

You believe your way, I'll believe mine, but every time my house shudders as a truck rumbles by, every time I hear thunder in the distance, every time I see a snowstorm blowing across the lake, or hear of another flood, I shiver and I wonder: Is this the end?

But it doesn't have to be that way.

Sure, like those fat, happy, and healthy woolly mammoths of yore, we may have a fatal flaw (we're too heavy and can't climb out of the snow), but we also have a saving grace.

We've been warned. They weren't.

We can think. We can plan. We can use our wits to survive. We can monitor our decreasing magnetic field. We can monitor temperatures in the seas. We can monitor seismic and volcanic activity. We can monitor winter storms.

We can, in short, prepare. They couldn't.

But first we must understand that it's not our fault. We must understand that it's part of a natural cycle. We must understand that we're not destroying our world; our world is about to destroy us. Only then will we ask the right questions. Only then will we find the right answers.

What can we do to prepare?

Perhaps we can seed the clouds above our oceans to make it rain over water instead of on land. Perhaps we can spread a kind of non-toxic oil on the seas to retard evaporation. Perhaps we can develop a kind of fish that will survive our warming seas.

Far-fetched? Maybe. The point is that we should consider every possible alternative.

Some things, though, we *must* do.

We must monitor our declining magnetic field. If the rate of decline quickens, or if, God help us, our magnetic field begins fluctuating wildly, we'll know that giant earthquakes and violent volcanic eruptions could soon ravage the earth.

For the same reason, we must monitor sea temperatures. If water temperatures suddenly shoot through the roof, we'll know that giant storms and massive floods are sure to follow.

We must also monitor underwater volcanism. But how can we, when only five percent of the ocean floor has ever been mapped in detail? We must therefore map the rest of it . . . and soon.

We must study the El Niño phenomenon for clues. As weather patterns change, where will it snow? Where will it flood? Which areas will undergo drought? Which will be swept by tsunami? Given our recent advances in technology, these should be relatively easy questions . . . and the answers could save millions of lives.

We must map the 11,500-year flood plain. Knowing which areas to avoid during the coming floods could also save millions of lives. Similarly, we must map the flood-prone areas along our coasts.

What else can we do?

We must find ways to protect our power transmission lines, our phone lines, our TVs, our radios, our computers, not only from the coming storms, floods, and tectonic terror, but also from the magnetic reversal itself.

We must implement the VAN method of predicting earthquakes in advance (by tracking electrotelluric currents in the soil). That, too, could save millions of lives.

And we must take a long, hard look at our business-bashing environmental policies. Every dollar we waste trying to fix something that we didn't cause and cannot change is one less dollar that could have been used to invent and build new tools of survival.

Then we must make decisions.

Should we all move south? Not necessarily. Watching a few billion people fight and die over the same few chunks of southern real estate would not be a pretty picture.

Besides, how far south is south enough?

Should we all move to the mountains? Think twice. If land rises into the sky in sync with equinoctial precession, our mountains could soon be hammered by the worst earthquakes in 11,500 years, or explode in gigantic volcanic eruptions.

Besides, if snow lines descend by six-tenths of a mile, as they did during the last ice age . . . well, you get the picture.

Should we keep well-stocked boats in our yards like modern-day Noahs? Should we keep snowshoes under our beds (to defeat our fatal flaw)? What about food? What will we eat if northern grainfields are covered with snow and southern crops covered with mud . . . or completely washed away by the worst floods in 11,500 years?

We need answers to questions that we don't even know to ask.

So let's start asking. Let's ask everyone—professors, students, scientists, business leaders, government leaders, NASA, NOAA, the USGS, *everyone*—to pull together in search of survival solutions.

And a little bit of luck wouldn't hurt.

Good luck.

BIBLIOGRAPHY

Alexander, Tom, 1975, "A revolution called plate tectonics has given us a whole new Earth," *Smithsonian*, p. 31-40, Jan 1975.

Alexander, Tom, 1975, "Plate tectonics has a lot to tell us about the present and future Earth," *Smithsonian*, p. 39-47, Feb 1975.

Allaby, Michael, and Lovelock, James, 1983, *The Great Extinction*.

Alvarez, L. W., Alvarez, W., Asaro, F. and Michel, H. V., 1980, "Extraterrestrial cause for the Cretaceous-Tertiary Extinction," *Science*, Vol. 208, p. 1095-1108.

Alvarez, Walter, 1986, "Toward a Theory of Impact Crises," *Eos*, Vol. 67, p. 649, 653-655, 658, 2 Sep 1986.

Alvarez, Walter, Alvarez, L. W., Asaro, F. and Michel, H. V., 1984, "The End of the Cretaceous: Sharp Boundary or Gradual Transition?" *Science*, Vol. 223, p. 1183-1186, 16 Mar 1984.

Alvarez, Walter, and Asaro, Frank, 1990, "An Extraterrestrial Impact," *Scientific American*, p. 78-84, Oct 1990.

Alvarez, Walter, and Muller, Richard A., 1984, "Evidence from crater ages for periodic impacts on the Earth," *Nature*, Vol. 308, p. 718-720, 19 Apr 1984.

Anderson, Roger Y., 1982, "A long geoclimatic record from the Permian," *Journal of Geophys. Res.*, Vol. 87, p. 7285-7294.

Anklin, M., et al., (Greenland Ice-core Project), 1993, "Climate instability during the last interglacial period recorded in the GRIP ice core," *Nature*, Vol. 364, p. 203-207, 15 Jul 1993.

Ardrey, Robert, 1961, *African Genesis*.

Armstrong, R. L., 1971, "Glacial erosion and the variable isotopic composition of strontium in sea water," *Nature Phys. Sci.*, Vol. 230, p. 132-133, 5 Apr 1971.

A. S. W. (initials only), 1903, "The New Mammoth at St. Petersburg," *Nature*, Vol. 68, p. 297-298, 30 Jul 1903.

Atwater, Tanya, 1970, "Implications of Plate Tectonics for the Cenozoic Tectonic Evolution of Western North America," *Geol. Soc. Am. Bull.*, Vol. 81, p. 3513-3536. (Also in *Plate Tectonics and Geomagnetic Reversals*, p. 583-609.)

Auzende, J. M., Sinton, J. M. and Scientific Party, 1994, "NAUDUR explorers discover recent volcanic activity along the East Pacific Rise, *Eos*, Vol. 75, p. 601, 604, 605, 20 Dec 1994.

Axelrod, D. I., 1981, "Role of volcanism in climate and evolution," *Geol. Soc. Am. Spec. Pap. 185*, p. 1-32.

Baadsgaard, H. and Lerbekmo, J. F., 1980, "A Rb/Sr age for the Cretaceous-Tertiary boundary (Z coal), Hell Creek, Montana," *Can. Journal of Earth Sci.*, Vol. 17, p. 671-673.

Bahcall, John N. and Bahcall, Safi, 1985, "The Sun's motion perpendicular to the galactic plane," *Nature*, Vol. 316, p. 706-708, 22 Aug 1985.

Bakker, Robert T., 1986, *The Dinosaur Heresies*.

Barbetti, M. and Flude, K., 1979, "Geomagnetic variation during the late Pleistocene period and changes in the radiocarbon time scale," *Nature*, Vol. 279, p. 202-205, 17 May 1979.

Barbetti, M., 1979, "Palaeomagnetic field strengths from sediments baked by lava flows of the Chaîne des Puys, France," *Nature*, V. 278, p. 153-156, 8 Mar 1979.

Bard, E., Hamelin, B., Fairbanks, R. G. and Zinder, A., 1990, "Calibration of the ^{14}C timescale over the past 30,000 years using mass spectrometric U-Th ages from Barbados Corals," *Nature*, Vol. 345, p. 405-410, 31 May 1990.

Barnola, J. M., Raynaud, D., Korotkevich, Y. S. and Lorius, C., 1987, "Vostok ice core provides 160,000 year record of atmospheric CO_2," *Nature*, Vol. 329, p. 408-413, 1 Oct 1987.

Barry, J. P., Baxter, C. H., Sagarin, R. D. and Gilman, S. E., 1995, "Climate-Related, Long-Term Faunal Changes in a California Rocky Intertidal Community," *Science,* Vol. 267, p. 672-675, 3 Feb 1995.

Beard, John H., 1969, "Pleistocene Paleotemperature Record based on Planktonic Foraminifers, Gulf of Mexico," *Trans. Gulf Coast Assoc. Geol. Soc.*, Vol. 19, p. 535-553.

Beer, Jürg, Oeschger, H., Andrée, M., Bonani, G., Suter, M., Wölfli, W. and Langway, C. C. Jr., 1984, "Temporal Variations in the ^{10}Be Concentration Levels Found in the Dye 3 Ice Core, Greenland," *Annals of Glaciology*, V. 5, p. 16-17.

Begley, Sharon, "This is Global Warming?" *Newsweek*, p. 20-27, 22 Jan 1996.

Beiser, Arthur, 1962, *The Earth*.

Benioff, Hugo, 1954, "Orogenesis and Deep Crustal Structure: Additional Evidence from Seismology," *Geol. Soc. Am. Bull.*, Vol. 65, p.385-400. (Also in *Plate Tectonics and Geomagnetic Reversals*, p. 295-310.)

Bentley, Charles R., 1993, "Antarctic Mass Balance and Sea Level Change," *Eos*, Vol. 74, p. 585-586, 14 Dec 1993.

Bentley, Charles R. and Giovinetto, M. B., 1992, "Mass balance of Antarctica and sea level change (abstract)," *Eos*, Vol. 203, p. 203.

Benton, Michael J., 1985, "Interpretations of Mass Extinctions," *Nature*, Vol. 314, p. 496-497, 11 Apr 1985.

Black, D. J., 1967, "Cosmic ray effects and faunal extinctions at geomagnetic field reversals," *Earth and Planet. Sci. Lett.*, Vol. 3, p. 225-236.

Bloom, A. L., Broecker, W. S., Chappell, J. M., Matthews, R. and Mesolella, K., 1974, "Quaternary sea-level fluctuations on a tectonic coast: New $^{230}Th/^{234}U$ dates from the Huon Peninsula, New Guinea," *Quat. Res.*, Vol. 4, p. 185-205.

Bogaard, Paul v.d. and Schmincke, Hans-Ulrich, 1985, "Laacher See Tephra: A widespread isochronous late Quaternary tephra layer in central and northern Europe," *Geol. Soc. Am. Bull.*, Vol. 96, p. 1554-1571, Dec 1985.

Bohor, B. F., Foord, E. E., Modreski, P. J. and Triplehorn, D. M., 1984, "Mineralogic Evidence for an Impact Event at the Cretaceous-Tertiary Boundary," *Science,* Vol. 224, p. 867-869, 25 May 1984.

Böhnel, H., Reismann, N., Jäger, G., Haverkamp, U., Negendank, J. F. W. and Schmincke, H. -U., 1987, "Paleomagnetic investigation of Quaternary West Eifel volcanics (Germany): indication for increased volcanic activity during geomagnetic excursion/event?" *Journal of Geophys.*, Vol. 62, p. 50-61.

Bolt, B. A., Horn, W. L., MacDonald, G. A. and Scott, R. F., 1976, *Geological Hazards*, p. 124-131.

Bourgeois, Joanne, Hansen, Thor A., Wiberg, Patricia L. and Kauffman, Erle G., 1988, "A Tsunami Deposit at the Cretaceous-Tertiary Boundary in Texas," *Science,* Vol. 241, p. 567-570, 29 Jul 1988.

Bradley, R. S., Diaz, H. F., Eischeid, J. K., Jones, P. D., Kelly, P. M. and Goodess, C. M., 1987, "Precipitation Fluctuations over Northern Hemisphere Land Areas Since the Mid-19th Century," *Science*, Vol. 237, p. 171-175, 10 Jul 1987.

Bradley, W. H., 1930, "The varves and climate of the Green River epoch," *U.S. Geol. Surv. Prof. Pap. 158-E*, p. 87-110, 1930.

Bray, J. R., 1977, "Pleistocene Volcanism and Glacial Initiation," *Science*, Vol. 197, p. 251-254, 15 Jul 1977.

Bretz, J Harlen, 1923, "The Channeled scabland of the Columbia Plateau, *Journal of Geology*, Vol. 31, p. 617-649, Nov-Dec 1923.

Bretz, J Harlen, 1969, "The Lake Missoula Floods and the Channeled Scabland," *Journal of Geology*, Vol. 77, p. 505-543.

Bretz, J Harlen, 1927, "Channeled scabland and the Spokane flood," in Proceedings of the Academy and Affiliated Societies, Geological Society, *Journal of the Washington Academy of Sciences*, Vol. 17, No. 8, p. 200-211.

Briden, J. C., 1976, "Application of palaeomagnetism to Proterozoic tectonics," *Phil. Trans. R. Soc. Lond.*, Series A., Vol. 280, p. 405-416.

Broad, William J. 1995, "Hot Vents in the Sea Floor May Drive El Niño," *The New York Times,* 25 Apr 1995.

Broecker, Wallace S., 1984, "Terminations," in *Milankovitch and Climate, Part 2*, A. L. Berger et al., editors, p. 687-698.

Broecker, Wallace S., 1994, "Massive iceberg discharges as triggers for global climate change," *Nature*, Vol. 372, p. 421-424, 1 Dec 1994.

Broecker, W., Andree, M., Wolfli, W., Oeschger, H., Bonani, G., Kennett, J. and Peteet, D., 1988, "The chronology of the last deglaciation: Implications to the cause of the Younger Dryas event," *Paleoceanography,* V. 3, p. 1-19, Feb 1988.

Broecker, Wallace S. and Denton, George H., 1990, "What Drives Glacial Cycles?" *Scientific American,* p. 48-56, Jan 1990.

Broecker, Wallace S., Ewing, M., and Heezen, B. C., 1960, "Evidence for an abrupt change in climate close to 11,000 years ago," *Am. Journal of Sci.,* Vol. 258, p. 429-448, Jun 1960.

Broecker, Wallace S. and Thurber, D. L., 1965, "Uranium-series dating of corals and oolites from Bahamian and Florida Key limestones," *Science,* Vol. 149, p. 58-60, 2 Jul 1965.

Broecker, Wallace S., Thurber, D. L., Goddard, J., Ku, Teh-lung, Matthews, R. K. and Mesolella, K. L., 1968, "Milankovitch Hypothesis Supported by Precise Dating of Coral Reefs and Deep-Sea Sediments," *Science*, Vol. 159, p. 297-300, 19 Jan 1968.

Bromley, R. G., 1979, "Chalk and bryozoan limestone: Facies, sediments, and depositional environments," In: *Cretaceous-Tertiary Boundary Events Symposium*, Eds. Birkelund and Bromley, p. 16-32.

Bucha, V., Taylor, R. E., Berger, R. and Haury, E. W., 1970, "Geomagnetic intensity: changes during the past 3000 years in the western hemisphere," *Science*, Vol. 168, p. 111-114, 3 Apr 1970.

Burton, Paul W., 1985, "Electrical Earthquake Prediction," *Nature*, Vol. 315, p. 370-371, 30 May 1985.

Byerlee, James D., Lockner, David A. and Johnson, Malcolm J., 1982, "Earthquake lights explained by frictional heating," *Earth Sciences,* fall 1982, p. 32-33.

Caldera, Ken and Kasting, James F., 1992, "Susceptibility of the early Earth to irreversible glaciation caused by carbon dioxide clouds," *Nature*, Vol. 359, p. 226-228, 17 Sep 1992.

Calder, Nigel, 1979, *Einstein's Universe*.

Campbell, I. H., Czamanske, G. K., Fedorenko, V. A., Hill, R. I. and Stepanov, V., 1992, "Synchronism of the Siberian Traps and the Permian-Triassic Boundary," *Science*, Vol. 258, p. 1760-1763, 11 Dec 1992.

Campsie, J., Johnson, G. L., Jones, J. E. and Rich, J. E., 1984, "Episodic Volcanism and evolutionary crises," *Eos*, Vol. 65, p. 796-800.

Canby, Thomas Y., 1990, "Earthquake—Prelude to the Big One?" *National Geographic*, p. 76-105, May 1990.

Canby, Thomas Y., 1973, "California's San Andreas Fault," *National Geographic*, p. 38-53, Jan 1973.

Cande, Steven C. and Kent, Dennis V., 1992, "A New Geomagnetic Polarity Time Scale for the Late Cretaceous and Cenozoic," *Journal of Geophys. Res.*, Vol 97, No. B10, p. 13,917-13,951, 10 Sep 1992.

Cande, Steven C. and Kent, Dennis V., 1995, "Revised calibration of the geomagnetic polarity timescale for the Late Cretaceous and Cenozoic," *Journal of Geophys. Res.*, Vol. 100, No. B4, p. 6093-6095, 10 Apr 1995.

Caputo, Mario V. and Crowell, John C., 1985, "Migration of glacial centers across Gondwana during Paleozoic Era," *Geol. Soc. Am. Bull.*, Vol. 96, p. 1020-1036, Aug 1985.

Carey, S. Warren, 1976, *The Expanding Earth*.

Carey, S. Warren, 1981, *The Expanding Earth, a symposium*.

Carey, S. Warren, 1989, *Theories of the Earth and Universe*.

Carmichael, C. M., 1967, "An Outline of the Intensity of the Paleomagnetic Field of the Earth," *Earth and Planet. Sci. Letters*, Vol. 3, p. 351-354. (Also in *Megacycles: Long-Term Episodicity in Earth and Planetary History*.)

Carrigan, Charles R. and Gubbins, David, 1979, "The Source of the Earth's Magnetic Field," *Scientific American*, p. 118-130, Feb 1979.

Champion, D. E., Lanphere, M. A. and Kuntz, M. A., 1988, "Evidence for a new geomagnetic reversal from Lava flows in Idaho: Discussion of short polarity reversals in the Brunhes and late Matuyama polarity chrons," *Journal of Geophys. Res.*, Vol. 93, p. 11667-11680, 10 Oct 1988.

Chappell, J., 1973, "Astronomical Theory of Climatic Change: Status and Problem," *Quaternary Research*, Vol. 3, p. 221-236.

Chappell, J., 1974, "Relationships between sealevels, ^{18}O variations and orbital perturbations, during the past 250,000 years," *Nature*, Vol. 252, p. 199-202, 15 Nov 1974.

Chappellaz, J., Barnola, J. M., Raynaud, D., Korotkevich, Y. S. and Lorius, C., 1990, "Ice-core record of atmospheric methane over the past 160,000 years," *Nature*, Vol. 345, p. 127-131, 10 May 1990.

Chauvin, Annick, Duncan, R. A., Bonhommet, N. and Levi, S., 1989, "Paleointensity of the Earth's magnetic field and K-Ar dating of the Louchardière volcanic flow (central France): new evidence for the Laschamp excursion," *Geophys. Res. Lett.*, Vol. 16, p. 1189-1192.

Chernicoff, Stanley and Venkatakrishnan, Ramesh, 1995, *Geology*.

Chorlton, Windsor, *Ice Ages*, Time-Life Books, 1983.

Christiansen, Robert L., 1984, "Yellowstone Magmatic Evolution: Its Bearing on Understanding Large Volume Explosive Volcanism," in *Explosive Volcanism: Inception, Evolution, and Hazards*, p. 84-95.

Christodoulidis, D. C., Smith, D. E., Kolenkiewicz, R., Klosko, S. M., Torrence, M. H. and Dunn, P. J., 1985, "Observing tectonic plate motions and deformation from a satellite laser ranging," *Journal of Geophys. Res.*, Vol. 90, p. 9249-9263, 30 Sep 1985.

Clark, David L., Cheng-Yuan, Wang, Orth, Charles J. and Gilmore, James S., 1986, "Conodont Survival and Low Iridium Abundances Across the Permian-Triassic Boundary in South China," *Science,* Vol. 233, p. 984-986, 29 Aug 1986.

Clark, H. C. and Kennett, J. P., 1973, "Paleomagnetic Excursion Recorded in Latest Pleistocene Deep-sea sediments, Gulf of Mexico," *Earth and Planet. Sci. Lett.* Vol. 19, p. 267-274.

Clark, Peter U. and Lea, Peter, D., editors, 1992, "The Last Interglacial Transition in North America," *Geol. Soc. Am. Spec. Pap. 270.*

Clemens, S. C., Farrell, J. W. and Gromet, L. P., 1993, "Synchronous changes in seawater strontium isotope composition and global climate," *Nature*, Vol. 363, p. 607-610, 17 Jun 1993.

Coe, Robert S., Gromme, S. and Mankinen, E. A., 1984, "Geomagnetic Paleointensities from Excursion Sequences in Lavas on Oahu, Hawaii," *Journal of Geophys. Res.*, Vol. 89, No. B2, p. 1059-1069, 10 Feb 1984.

Coe, Robert S. and Prévot, Michel, 1989, "Evidence suggesting extremely rapid field variations during a geomagnetic reversal," *Earth and Planet. Sci. Lett.,* Vol. 92, p. 292-298.

Coe, Robert S., Prévot, Michel, and Camps, P., 1995, "New Evidence for extraordinarily rapid change of the geomagnetic field during a reversal," *Nature*, Vol. 374, p. 687-692, 20 Apr 1995.

Colbert, Edwin H., 1968, *The Great Dinosaur Hunters and Their Discoveries.*

Collinson, David W., 1983, *Methods in rock magnetism and palaeomagnetism: techniques and instrumentation.*

Converse, D. R., Holland, H. D. and Edmond, J. M., 1984, "Flow rates in the axial hot springs of the East Pacific Rise (21°N): implications for the heat budget and the formation of massive sulfide deposits," *Earth and Planetary Sci. Lett.*, Vol. 69, p. 159-175.

Corliss, William R., 1978, *Geological Facts.*

Courtillot, Vincent E. and Besse, Jean, 1986, "Paleographic Evolution of the Tethys and Indian Ocean bordering Continents since the Breakup," *Eos,* Vol. 67, p. 925, 4 Nov 1986.

Courtillot, Vincent E. and Besse, Jean, 1987, "Magnetic Field Reversals, Polar Wander, and Core-Mantle Coupling," *Science,* Vol. 237, p. 1140-1147, 4 Sep 1987.

Courtillot, Vincent E., 1990, "A Volcanic Eruption," *Scientific American*, p. 85-92, Oct 1990.

Cox, Allan, 1969, "Geomagnetic reversals," *Science*, Vol. 163, p. 237-245, 17 Jan 1969. (Also in *Plate Tectonics and Geomagnetic Reversals*, p. 207-220.)

Cox, Allan, 1973, *Plate Tectonics and Geomagnetic Reversals.*

Cox, Allan, Dalrymple, G. Brent, and Doell, R. R., 1967, "Reversals of the Earth's Magnetic Field," *Scientific American*, Vol. 216, p. 44-54. (Also in *Plate Tectonics and Geomagnetic Reversals*, p. 188-206.)

Cox, Allan, Doell, Richard R. and Dalrymple, G. Brent, 1963, "Geomagnetic polarity epochs and Pleistocene geochronometry," *Nature*, Vol. 198, p. 1049-1051.

Cox, Allan, Doell, Richard R. and Dalrymple, G. Brent, 1964, "Reversals of the Earth's Magnetic Field," *Science*, Vol. 144, p. 1537-1543. (Also in *Plate Tectonics and Geomagnetic Reversals*, p. 169-178.)

Cox, Allan and Hart, Robert Brian, 1986, *Plate Tectonics, How it Works.*

Crain, I. K., 1971, "Possible Direct Causal Relation between Geomagnetic Reversals and Biological Extinctions," *Geol. Soc. Amer. Bull.*, Vol. 82, p. 2603-2606.

Crain, I. K. and Crain, P. L., 1970, "New stochastic model for geomagnetic reversals," *Nature*, Vol. 228, p. 39-41.

Crain, I. K., Crain, P. L. and Plant, M. G., 1969, "Long Period Fourier spectrum of geomagnetic reversals," *Nature*, Vol. 223, p. 283.

Creer, Kenneth M., 1975, "On a tentative correlation between changes in the geomagnetic polarity bias and reversal frequency and the Earth's rotation through Phanerozoic time," in *Growth Rhythms and the History of the Earth's Rotation*, G. D. Rosenberg and S. K. Runcorn, editors, p. 293-318.

Croll, James, 1875, *Climate and Time.*

Crowell, John C., 1977, *Problems concerning the Late Paleozoic glaciation of Gondwanaland: IV International Gondwana Symposium, Calcutta, Sec V., Glacial Deposits*, p. 347-351.

Crowley, Thomas J., 1983, "The Geological Record of Climatic Change," *Rev. Geophys. Space Phys.*, Vol. 21, p. 828-875.

Crowley, Thomas J. and North, Gerald R., 1988, "Abrupt Climate Change and Extinction Events in Earth History, *Science*, Vol. 240, p. 996-1002, 20 May 1988.

Crutzen, P. J., Isaksen, I. and Reid, G. C., 1975, "Solar Proton Events: Stratospheric Sources of Nitric Oxide," *Science*, Vol. 189, p. 457-459, 8 Aug 1975.

Cuppy, Will, 1941, *How to Become Extinct.*

Damon, P. E., 1969, "Climatic versus magnetic perturbation of the atmospheric carbon 14 reservoir," in Nobel Symposium - 12th: 1969, *Radiocarbon Variations and Absolute Chronology: Proceedings*, Ingrid U. Olsson editor (Wiley Interscience.)

Damon, P. E., 1971, "The Relationship between Late Cenozoic Volcanism and Tectonism and Orogenic-Epeirogenic Periodicity," in *The Late Cenozoic Glacial Ages*, Karl K. Turekian, ed. p. 15-23 and 31-33. (Also in *Megacycles: Long-Term Episodicity in Earth and Planetary History*.)

Dana, James D., 1894, *Manual of Geology*, p. 1007.

Dansgaard, W., 1981, "Ice Core studies: Dating the past to find the future," *Nature*, Vol. 290, p. 360-361, 2 Apr 1981.

Dansgaard, W. and Duplessy, J., 1981, "The Eemian interglacial and its termination," *Boreas*, Vol. 10, p. 219-228.

Dansgaard, W., Johnsen, S. J., Clausen, H. B., Dahl-Jensen, D., Gundestrup, N. S., Hammer, C. U., Hvidberg, C. S., Steffensen, J. P. Sveinbjörnsdottir, A. E., Jouzel, J. and Bond, G., 1993, "Evidence for general instability of past climate from a 250-kyr ice-core record," *Nature*, Vol. 364, p. 218-220, 15 Jul 1993.

Dansgaard, W., Johnsen, S. J., Clausen, H. B. and Langway, C. C. Jr., 1971, "Climatic Record Revealed by Camp Century Ice Core," in *Late Cenozoic Glacial Ages*, Karl K. Turekian, editor, p. 37-56.

Dansgaard, W., White, J. W. C. and Johnsen, S. J., 1989, "The abrupt termination of the Younger Dryas climate event," *Nature*, Vol. 339, p. 532-534, 15 Jun 1989.

Davis, Marc, Hut, Piet, and Muller, Richard A., 1984, "Extinction of Species by periodic comet showers," *Nature*, Vol. 308, p. 715-717, 19 Apr 1984.

Dawson, Alastair G., 1992, *Ice age earth: late quaternary and climate.*

Dearnley, R., 1965, "Orogenic Fold-Belts and Continental Drift," *Nature*, Vol. 206, p. 1083-1087, 12 Jun 1965.

Dearnley, R., 1966, "Fold-belts and a Hypothesis of Earth Evolution," *Phys. and Chem. Earth*, Vol. 7, p. 1-114. (Also in *Megacycles: Long-Term Episodicity in Earth and Planetary History*.)

de Boer, P. L. and Shackleton, N., 1984, "Pre-Pleistocene Evidence of Orbital Forcing," in *Milankovitch and Climate, Part 2*, A. L. Berger et al., editors.

Decker, Robert, and Decker, Barbara, 1981, *Volcanoes.*

Denham, C. R. and Cox, Allan V., 1971, "Evidence that the Laschamps polarity event did not occur 13,300-30,400 years ago," *Earth and Planet. Sci. Lett.*, Vol. 13, p. 181-190.

Derr, John S., 1973, "Earthquake Lights: A review of Observations and Present Theories," *Bull. Seismol. Soc. Am.*, Vol. 63, no. 6, p. 2177-2187, Dec 1973.

Dewey, J. F. and Bird, John M., 1970, "Mountain Belts and the New Global Tectonics," *Journal of Geophys. Res.*, Vol. 75, p. 2625-2647. (Also in *Plate Tectonics and Geomagnetic Reversals*, p. 610-631.)

Dewey, J. F. and Horsfield, Brenda, 1970, "Plate Tectonics, orogeny and continental growth," *Nature*, Vol. 225, p. 521-525.

Doake, Christopher S. M., 1978, "Climatic change and geomagnetic field reversals: A statistical correlation," *Earth and Planet. Sci. Lett.*, Vol. 38, p. 313-318.

Doell, Richard R. and Dalrymple, G. Brent, 1966, "Geomagnetic Polarity Epochs: A New Polarity Event and the Age of the Brunhes-Matuyama Boundary," *Science*, Vol. 152, p. 1060-1061. (Also in *Tectonics and Geomagnetic Reversals*, p. 179-181.)

Donaldson, J. A., McGlynn, J. C., Irving, E. and Park, J. K., 1973, "Drift of the Canadian Shield," in Nato Advanced Studies Institute *Implications of Continental Drift to the Earth Sciences*, Vol. 1 (D. H. Tarling and S. K. Runcorn, editors), p. 3-17.

Donovan, Stephen K., 1987, "Iridium anomalous no longer?" *Nature*, Vol. 326, p. 331-332, 26 Mar 1987.

Dott, R. H., 1969, "Circum-Pacific Late Cenozoic Structural Rejuvenation: Implications for Sea Floor Spreading," *Science*, Vol. 166, p. 874-876.

Douglas, John et al., 1989, "A Storm From the Sun," *EPRI Journal* [Electrical Power Research Institute], p. 14-21, July/Aug 1989.

Egyed, L., 1956, Letter to *Nature*, Vol. 178, p. 534, 8 Sep 1956.

Eldredge, Niles, 1985, *Time Frames.*

Eldredge, Niles and Gould, Stephen Jay, 1972, *Punctuated Equilibria: An Alternative to Phyletic Gradualism.*

Elliott, David K., editor, *Dynamics of Extinction.*

Elsasser, Walter, Ney, E. P. and Winckler, J. R., 1956, "Cosmic-ray intensity and geomagnetism," *Nature*, Vol. 178, p. 1226-1227, 1 Dec 1956.

Emiliani, Cesare, 1980, "Death and Renovation at the End of the Mesozoic," *Eos*, Vol. 61, p. 505-506, 24 Jun 1980.

Emiliani, Cesare et al., 1975, "Paleoclimatological Analysis of Late Quaternary Cores from the Northeastern Gulf of Mexico," *Science*, Vol. 189, 26 Sep 1975.

Engel, A. E. Jr., Itson, S. P., Engel, C. G., Stickney, D. M. and Cray, E. J., 1974, "Crustal evolution and global tectonics: a petrogenic view," *Geol. Soc. Am. Bull. 85*, p. 843-858.

Erickson, Jon, 1990, *Ice Ages, Past and Future*.

Erwin, Douglas H., 1994, "The Permo-Triassic extinctions," *Nature*, Vol. 367, p. 231-236, 20 Jan 1994.

Evernden, J. F. and Kistler, R. W., 1970, "Chronology of emplacement of Mesozoic batholithic complexes in California and western Nevada," *U.S. Geol. Survey Prof. Pap. 623*. 42 pp.

Ewing, J. and Ewing, M., 1967, "Sediment Distribution on the Mid-Ocean Ridges with Respect to Spreading of the Sea Floor," *Science*, Vol. 156, p. 1590-1592.

Fabricius, F. H., Friedrichsen, V. and Jacobshagen, V., 1970, "Paläotemperaturen and Palaoklima in Obertrias und Lias der Alpen," *Geologische Rundschau*, Vol. 59, p. 805-826.

Fairbridge, Rhodes W., 1977, "Global climate change during the 13,500 B.P. Gothenburg geomagnetic excursion," *Nature*, Vol. 265, p. 430-431, 3 Feb 1977.

Farrand, William R., 1961, "Frozen Mammoths and Modern Geology," *Science*, Vol. 133, p. 729-735, 17 Mar 1961.

Findley, Rowe, 1981, "St. Helens, Mountain with a Death Wish," *National Geographic*, p. 2-65, Jan 1981.

Fischer, A. G., 1980, "Gilbert-Bedding rhythms and geochronology," *Geol. Soc. Am. Spec. Pap 183*, p. 93-104.

Fischer, A. G., 1981, "Climatic oscillations in the biosphere," *Biotic Crises in Ecological and Evolutionary Time*, Nitecki, M. H., editor, p. 103-131.

Fitch, F. J. and Miller, J. A., 1965, "Major Cycles in the History of the Earth," *Nature*, Vol. 206, p. 1023-1027. (Also in *Megacycles: Long-Term Episodicity in Earth and Planetary History*.)

Fodor, R. V., 1981, *Frozen Earth: Explaining the Ice Ages*.

Forbes, William T. M., 1931, "The Great glacial cycle," *Science*, Vol. 74, p. 294-295, 18 Sep 1931.

Force, E. R., 1984, "A Relation Among Geomagnetic Reversals, Seafloor Spreading Rate, Paleoclimate, and Black Shales," *Eos*, Vol. 65, p. 18-19, 17 Jan 1984.

Forsyth, D. and Uyeda, S., 1975, "On the Relative Importance of the Driving Forces of Plate Motion," *Geophys. J. R. Astr. Soc.*, Vol. 43, p. 163-200.

Froelich, Philip N., 1993, "Ruling in the improbable," *Nature*, Vol. 363, p. 585-587, 17 Jun 1993.

Gaffin, Stuart, 1987, "Phase difference between sea level and magnetic reversal rate," *Nature*, Vol. 329, p. 816-819, 29 Oct 1987.

Gantenbein, Douglas, 1995, "El Niño the Weathermaker," *Popular Science*, p. 76-82, May 1995.

Gates, Bill, *Seattle Post-Intelligencer*, 13 Mar 1996.

Gillot, P. Y., Labeyrie, J., Laj, C., Valladas, G., Guerin, G., Poupeau, G. and Delibrias, G., 1979, "Age of the Laschamp paleomagnetic excursion revisited," *Earth and Planet. Sci. Lett.*, Vol. 42, p. 444-450, 1979.

Gladkih, Mikhail I., Kornietz, Ninelj L. and Soffer, Olga, 1984, "Mammoth-Bone Dwellings on the Russian Plain," *Scientific American*, Vol. 250, p. 164-175, Nov 1984.

Glass, B. P., Ericson, D. B., Heezen, B. B., Opdyke, N. D. and Glass, J. A., 1967, "Geomagnetic reversals and Pleistocene chronology," *Nature*, Vol. 216, p. 437-442.

Glass, B. P., Swinckl, M. B. and Zwart, P. A., 1979, "Australasian, Ivory Coast and North American tektite strewnfields: Size, mass and correlation with geomagnetic reversals and other earth events," *Proc. 10th Lunar Planet. Sci. Conf.*, p. 2535-2545.

Gokhberg, M. B., Morgonunov, V. A., Yoshino, T. and Timizawa, I., 1982, "Experimental Measurement of Electromagnetic Emissions Possibly Related to Earthquakes in Japan," *Journal of Geophys. Res.*, Vol. 87, p. 7824-7828, 10 Sep 1982.

Goldhammer, R. K., Dunn, P. A. and Hardie, L. A., 1987, "High frequency glacioeustasy sea-level oscillations and Milankovitch characteristics recorded in Middle Triassic cyclic platform carbonates, northern Italy," *Am. Journal of Sci.*, Vol. 287, p. 853-892, Nov 1987.

Gore, Rick, 1989, "What caused the earth's great dyings?" *National Geographic*, Vol. 175, p. 662-699, Jun 1989.

Gore, Rick, 1993, "Dinosaurs," *National Geographic*, Vol. 183, p. 2-53, Jan 1993.

Gould, Stephen Jay, 1977, "The Telltale Wishbone," *Natural History*, Vol. 86, No. 9, p. 26-36, Nov 1977.

Gould, Stephen Jay, 1980, *The Panda's Thumb*.

Gould, Stephen Jay, 1989, *Wonderful Life*.

Gradstein, Felix M. *et al.*, 1994, "A Mesozoic time scale," *Journal of Geophys. Res.*, Vol. 99, No. B12, p. 24,051-24,074, 10 Dec 1994.

Graham, Nicholas E., 1995, "Simulation of Recent Global Temperature Trends," *Science*, Vol. 267, p. 666-671, 3 Feb 1995.

Grayson, Donald K., 1996, "Ice Age Mammal Extinctions," A talk to the Northwest Paleontological Association, Seattle, Washington, 9 Mar 1996.

Gribben, John R. and Plagemann, Stephen H., 1975, *The Jupiter Effect*.

Grieve, R. A. F. and Dence, M. R., 1979, "The Terrestrial Cratering Record II. The Crater Production Rate," *Icarus*, Vol. 38, p. 230-242.

Grieve, R. A. F. and Robertson, P. B., 1979, "The Terrestrial Record I. Current Status of Observations," *Icarus*, Vol. 38, p. 212-229.

Grieve, R. A. F. and Sharpton, V. L., 1981, "The Cretaceous Tertiary extinction event: a cosmic catastrophe?" *Geos*, Vol. 10, p. 7-9.

GRIP (Greenland Ice-core Project), 1993, "Climate instability during the last interglacial period recorded in the GRIP ice core," *Nature*, Vol. 364, p. 203-207, 15 Jul 1993.

Grotzinger, J. P., 1986, "Upward Shallowing Platform Cycles: A Response to 2.2 billion years of Low-Amplitude, High-Frequency (Milankovitch Band) Sea Level Oscillations," *Paleoceanography*, Vol. 1, No. 4, p. 403-416, Dec 1986.

Grove, Noel, 1992, "Volcanoes: Crucibles of Creation," *National Geographic*, p. 5-41, Dec 1992.

Hall, C. M. and York, D., 1978, "K-Ar and 40Ar/39Ar age of the Laschamp geomagnetic polarity reversal," *Nature*, Vol. 274, p. 462-464, 3 Aug 1978.

Hallam, Anthony, 1971, "Re-evaluation of the Palaeogeographic Argument for an Expanding Earth," *Nature*, Vol. 232, p. 180-182.

Hallam, Anthony, 1977, "Secular changes in marine inundation of USSR and North America through the Phanerozoic," *Nature*, Vol. 269, p 769-772. (Also in *Megacycles: Long-Term Episodicity in Earth and Planetary History*.)

Hallam, Anthony, 1984, "Pre-Quaternary Sea Level Changes," *Annual Review Earth Planet Sci.* Vol. 12, p. 205-243.

Hallam, Anthony, 1984, "The causes of Mass Extinctions," *Nature* Vol. 308, 19 Apr 1984.

Hallam, Anthony, 1987, "End-Cretaceous Mass Extinction Event: Argument for Terrestrial Causation," *Science*, Vol. 238, p. 1237-1242, 27 Nov 1987.

Hallam, Anthony, 1990, "The end-Triassic mass extinction event," *Geol. Soc. Am. Spec. Pap. 247*.

Hallam, Anthony, 1992, *Phanerozoic Sea-Level Changes*.

Hardie, Lawrence A., Bossellini, Alfonso, and Goldhammer, Robert K., 1986, "Repeated Subaerial Exposure of Subtidal Carbonate Platforms, Triassic Northern Italy: Evidence for High Frequency Sea level Oscillations on a 10^4 Year scale," *Paleoceanography*, Vol. 1, No. 4, p. 447-457, Dec 1986.

Harland, W. B., Armstrong, R. L., Cox, A. V., Craig, L. E., Smith, A. G. and Smith, D. G., 1989, *A Geologic Time Scale 1989*.

Harrison, C. G. A., 1968, "Evolutionary Processes and reversals of the earth's magnetic field," *Nature*, Vol. 217, p. 46-47.

Harrison, C. G. A. and Funnel, B. M., 1964, "Relationship of Palaeomagnetic Reversals and Micropalaeontology in Two Late Caenozoic (*sic*) Cores from the Pacific Ocean," *Nature*, Vol. 204, p. 566, 7 Nov 1964.

Harrison, C. G. A. and Prospero, J. M., 1974, "Reversals of the Earth's magnetic field and climatic changes," *Nature*, Vol. 250, p. 563-565, 16 Aug 1974.

Hartman, William K., 1974, "Martian and Terrestrial Paleoclimatology: Relevance of Solar Variability," *Icarus*, Vol. 22, p. 301-311. (Also in *Megacycles: Long-Term Episodicity in Earth and Planetary History*.)

Harwood, J. M. and Malin, S. R. C., 1976, "Present trends in the Earth's magnetic field," *Nature*, Vol. 259, p. 469-471. 12 Feb 1976.

Hatfield, Craig B. and Camp, Mark J., 1970, "Mass Extinctions Correlated with Periodic Galactic Events," *Geol. Soc. Am. Bull.*, Vol. 81, p. 911-914, Mar 1970.

Hays, James D., 1970, "Stratigraphy and Evolutionary Trends of Radiolaria in North Pacific Deep-Sea Sediments," *Geol. Soc. Am.* Memoir 126, p. 185-218.

Hays, James D., 1971, "Faunal Extinctions and Reversals of the Earth's Magnetic Field," *Geol. Soc. Am. Bull.*, Vol. 82, p. 2433-2447, Sep 1971.

Hays, James D., 1973, "The Ice Age Cometh," *Saturday Review of the Sciences*, Vol. 1, p. 29-32, Apr 1973.

Hays, James D., Imbrie, John, and Shackleton, N. J., 1976, "Variations in the Earth's Orbit; Pacemaker of the Ice Ages," *Science*, Vol. 194, p. 1121-1132, 10 Dec 1976.

Hays, James D. and Opdyke, Neil D., 1967, "Antarctic Radiolaria, Magnetic Reversals, and Climatic Change," *Science,* Vol. 158, p. 1001-1011, 24 Nov 1967.

Hays, James D. and Pitman, Walter C. III, 1973, "Lithospheric Plate Motion, Sea Level Changes and Climatic and Ecological Consequences," *Nature,* Vol. 246, p. 18-22, 2 Nov 1973.

Hays, James D., Saito, Tsunemasa, Opdyke, Neil D. and Burckle, Lloyd H., 1969, "Pliocene-Pleistocene Sediments of the Equatorial Pacific: Their Paleomagnetic, Biostratigraphic, and Climatic Record," *Geol. Soc. Am. Bull.,* Vol. 80, p. 1481-1514, Aug 1969.

Heath, G. R., 1969, "Mineralogy of Cenozoic Deep-Sea Sediments from the Equatorial Pacific Ocean," *Geol. Soc. Am. Bull.,* Vol. 80, p. 1997-2018.

Heirtzler, James R., Dickson, G. O., Herron, Ellen M., Pitman, Walter C. III, Herron, M. M. and Langway, C. C., 1982, "Chloride, nitrate, and sulphate in the Dye 3 and Camp Century, Greenland ice cores" (abstract), *Eos,* Vol. 63, p. 298, 4 May 1982.

Herron, M. M. and Langway, C. C., 1982, "Chloride, nitrate, and sulphate in the Dye 3 and Camp Century, Greenland ice cores," *Eos,* Vol. 63, p. 298, 4 May 1982.

Hess, Harry H., 1962, "History of Ocean Basins," *Petrological Studies: A Volume in Honor of A. F. Buddington,* Engell, James and Leonard, eds, p. 599-620. Also in *Plate Tectonics and Geomagnetic Reversals,* p. 23-38. Also in *Adventures in Earth History,* Preston Cloud, ed., 1970, p. 277-292.)

Hibben, Frank C., 1946, *Lost Americans.*

Hickey, Leo J., 1981, "Land plant evidence compatible with gradual, not catastrophic, change at the end of the Cretaceous," *Nature,* Vol. 292, 6 Aug 1981, p. 529-531.

Hildebrand, Alan R. and Boynton, William V., 1990, "Proximal Cretaceous-Tertiary Boundary Impact Deposits in the Caribbean," *Science,* Vol. 248, p. 843-847, 18 May 1990.

Hildebrand, Alan R. and Boynton, William V., 1991, "Cretaceous Ground Zero," *Natural History,* Jun 91.

Hilgen, F. J., 1991, "Extension of the astronomically calibrated (polarity) time scale to the Miocene/Pliocene boundary, *Earth Planet. Sci. Lett.,* Vol. 107, p. 349-368, 1991.

Hill, Mary, "The Earth Speaks Softly," *Earth Science,* summer 1981, p. 30-31.

Hoffman, Kenneth A., 1988, "Ancient Magnetic Reversals: Clues to the Geodynamo," *Scientific American,* p. 76-83, May 1988.

Holmes, Arthur, 1944, "The Machinery of Continental Drift: The Search for a Mechanism," *Principles of Geology,* p. 505-509. (Also in *Plate Tectonics and Geomagnetic Reversals,* p. 19-22.)

Holser, William T., 1977, "Catastrophic chemical events in the history of the ocean," *Nature,* Vol. 267, p. 403-408. (Also in *Megacycles: Long-Term Episodicity in Earth and Planetary History.*)

Horner, John R., 1988, *Digging Dinosaurs.*

Hoyle, Fred, 1981, *Ice, the ultimate human catastrophe.*

Hsü, Kenneth J., 1986, *The Great Dying.*

Hsü, K. J., He, McKenzie, Weissert, Perch-Nielsen, Oberhänsli, Kelts, LaBrecque, Tauxe, Krähenbühl, Percival, Wright, Karpoff, Petersen, Tucker, Poore, Gombos, Pisciotto, Carman, Schreiber, 1982, "Mass Mortality and Its Environmental-and Evolutionary Consequences," *Science*, Vol. 216, p. 249-256, 16 Apr 1982.

Hsü, Kenneth J., McKenzie, J. A. and He, Q. X., 1982, "Terminal Cretaceous environmental and evolutionary changes, *Geol. Soc. Am. Spec. Pap. 190*, p. 317-328.

Hsü, Kenneth J., Montadert, L., Bernoulli, D., Cita, Maria Bianca, Erickson, A., Garrison, R. E., Kidd, R. B., Mèlierés, F. Müller, C. and Wright, Ramil, 1977, "History of the Mediterranean salinity crisis," *Nature*, Vol. 267, p. 399-403, 2 Jun 1977.

Hughes, D. W., 1979, "Earth's Cratering Rate," *Nature*, Vol. 281, p. 11, 6 Sep 1979.

Huyghe, Patrick, 1982, "Earthquakes: The Solar Connection," *Science Digest*, Vol. 90, p. 72-75 and 103, Oct 1982.

"Ice Ages Attributed to Orbit Changes," *Science News*, V. 110, p 356, 4 Dec 1976.

Imbrie, John and Imbrie, John Z., 1980, "Modeling the climatic response to orbital variations," *Science*, Vol. 207, p. 943-953.

Imbrie, John and Imbrie, Katherine Palmer, 1979, *Ice Ages: Solving the Mystery*.

Imbrie, John, Hays, J. D., Martinson, D. G, McIntyre, A., Mix, A. C., Morley, J. J., Pisias, N. G., Prell, W. L. and Shackleton, N. J., 1984, "The Orbital theory of the marine $\delta^{18}O$ record." (In *Milankovitch and Climate, Part 1*, A. L. Berger et al., editors, p. 269-305.

Irving, E. and Park, J. K., 1972, "Hairpins and Superintervals," *Canadian Journal of Earth Sci.* Vol. 9, p. 1318-1324. (Also in *Megacycles: Long-Term Episodicity in Earth and Planetary History*.)

Irving, E. and Pullaiah, G., 1976, "Reversals of the Geomagnetic Field, Magnetostratigraphy, and Relative Magnitude of Paleosecular Variations in the Phanerozoic," *Earth-Sci. Rev.* Vol. 12, p. 35-38 and 53-64. (Also in *Megacycles: Long-Term Episodicity in Earth and Planetary History*.)

Irving, E. and Robertson, W. A., 1969, "Test for polar wandering and some possible implications," *Journal of Geophys. Res.*, Vol. 74(4), p. 1026-1036.

Irving, Tony, personal communication.

Isacks, Bryan, and Molnar, Peter, 1969, "Mantle Earthquake Mechanisms and the Sinking of the Lithosphere," *Nature*, Vol. 223, p. 1121-1124. (Also in *Plate Tectonics and Geomagnetic Reversals*, p. 401-406.)

Isacks, Bryan, Oliver, Jack, and Sykes, Lynn R., "Seismology and the New Global Tectonics," *Journal of Geophys. Res.*, Vol. 73, p. 5855-5899. (Also in *Plate Tectonics and Geomagnetic Reversals*, p. 358-400.)

Jéhanno, C., Boclet, D., Froget, L., Lambert, B., Robin, E., Rocchia, R. and Turpin, L., 1992, "The Cretaceous-Tertiary boundary at Beloc, Haiti: No evidence for an impact in the Caribbean Area," *Earth Planet. Sci. Lett.*, Vol. 109, p. 229-241.

Johnson, J. G., 1971, "Timing and Coordination of Orogenic, Epeirogenic, and Eustatic Events," *Geol. Soc. Am. Bull.*, Vol. 82, p. 3263-3298.

Johnson, J. G., 1972, "Antler effect equals Haug effect," *Geol. Soc. Am. Bull.*, Vol. 83, p. 2497-2498.

Katz, H. R., 1971, "Continental margin in Chile - is tectonic style compressional or extensional?" *Am. Assoc. Geol. Bull.* Vol. 55, p. 1753-1758.

Keller, G. and Barrera, E., 1990, "The Cretaceous/Tertiary boundary impact hypothesis and the paleontological record," *Geol. Soc. Am. Spec. Pap. 247.*

Kennett, James P., 1977, "Cenozoic evolution of Antarctic glaciation, the Circum-Antarctic Ocean, and their impact on global paleooceanography," *Journal of Geophys. Res.,* Vol. 82, p. 3843-3860, 20 Sep 1977.

Kennett, James P., McBirney, A. R. and Thunell, Robert C., 1977, "Episodes of Cenozoic volcanism in the circum-Pacific region," *Journal of Volcanology and Geothermal Research,* Vol. 2, p. 145-163.

Kennett, James P. and Thunell, Robert C., 1975, "Global Increase in Quaternary Explosive Volcanism," *Science,* Vol. 187, p. 497-503, 14 Feb 1975.

Kennett, James P. and Thunell, Robert C., 1977, "Comments on Cenozoic Explosive Volcanism Related to East and Southeast Asian Arcs," in "Island Arcs, deep sea trenches, and back-arc basins," Talwani and Pitman, editors, *American Geophysical Union Maurice Ewing Series,* No. 1, p. 348-352.

Kennett, James P. and Watkins, N. D., 1970, "Geomagnetic polarity change, volcanic maxima and faunal extinction in the south Pacific," *Nature,* Vol. 227, p. 930-934, 29 Aug 1970.

Kent, Dennis V., 1981, "Asteroid extinction hypothesis," *Science,* Vol. 211, p. 649-650.

Kent, Dennis V. and Gradstein, Felix M., 1986, "A Jurassic to recent chronology," in *The Geology of North America,* Vol. M, *The Western North Atlantic Region,* edited by P. R. Vogt and B. E. Tucholke, p. 45-50.

Kent, Dennis V. and Opdyke, N. D., 1976, "Relative paleomagnetic field intensity variations from deep-sea and sediment cores," *Abstract Transactions Am. Geophys. Union,* Vol. 57, p. 237, 1976.

Kent, Dennis V. and Opdyke, N. D., 1977, "Paleomagnetic field intensity variation recorded in a Brunhes Epoch deep-sea sediment core," *Nature,* Vol. 266, p. 156-159, 10 Mar 1977.

Kerr, Richard A., 1980, "Quake Prediction by Animals Gaining Respect," *Science,* Vol. 208, p. 695-696, 16 May 1980.

Kerr, Richard A., 1984, "An Impact but No Volcano," *Science,* Vol. 224, p. 258, 25 May 1984.

Kerr, Richard A., 1987, "Milankovitch Climate Cycle Through the Ages," *Science,* Vol. 235, p. 973-974, 7 Feb 1987.

Kerr, Richard A., 1988, "Huge Impact is Favored K-T Boundary Killer," *Science,* Vol. 242, p. 865-867, 11 Nov 1988.

Kerr, Richard A., 1991, "Dinosaurs and Friends Snuffed Out?" *Science,* Vol. 251, p. 160-162, 11 Jan 1991.

Kerr, Richard A., 1992, "A Revisionist Timetable for the Ice Ages," *Science,* Vol. 258, p. 220-221, 9 Oct 1992.

Kerr, Richard A., 1992, "When Climate Twitches, Evolution Takes Great Leaps," *Science,* Vol. 257, p. 1622-1624, 18 Sep 1992.

Kerr, Richard A., 1993, "A Bigger Death Knell for the Dinosaurs?" *Science,* Vol. 261, p. 1518-1519, 17 Sept 1993.

Kerr, Richard A., 1993, "The Greatest Extinction Gets Greater," *Science,* Vol. 262, p. 1370-1371, 26 Nov 1993.

King, Chi-Yu, 1983, "Electromagnetic Emissions Before Earthquakes," *Nature,* Vol. 301, p. 377, 3 Feb 1983.

Kistler, R. W., Evernden, J. F. and Shaw, H. R., 1971, "Sierra Nevada plutonic cycle: Part I, origin of composite granitic batholiths," *Geol. Soc. Am. Bull.,* Vol. 82, p. 853-868, Apr 1971.

Kopper, John S., 1976, "Dating and Interpretation of Archeological Cave Deposits by the Paleomagnetic Method," Doctoral Thesis, Columbia University.

Krige, L. J., 1929, "Magmatic Cycles, Continental Drift and Ice Ages," *Proceedings of the Geological Society of South Africa,* p. 21-40.

Kukla, George J., 1975, "Missing link between Milankovitch and climate," *Nature,* Vol. 253, p. 600-603, 20 Feb 1975.

Kukla, George J., 1977, "Pleistocene Land-Sea Correlations I. Europe," *Earth-Science Reviews,* Vol. 13, p. 307-374.

Kukla, George J., Berger, A., Lotti, R. and Brown, J., 1981, "Orbital signature of interglacials," *Nature,* Vol. 290, p. 295-300, 26 Mar 1981.

Kukla, George J., and Gavin, J., 1992, "Insolation Regime of the Warm to Cold Transition," in *Start of a Glacial,* NATO ASI Series, Vol. 13, Kukla and Went editors.

Kukla, George J. and Matthews, Robley K., 1972, "When will the present interglacial end?" *Science,* Vol. 178, p. 190-191, 13 Oct 1972.

Kukla, George J., Matthews, R. K. and Mitchel, J. M. Jr., 1972, "The end of the Present Interglacial," *Quat. Res.,* Vol. 2, p. 261-269.

Kukla, George J. and Went, H., eds, 1992, *Start of a Glacial,* NATO ASI Series.

Kumar, Arun, Leetmaa, Ants, and Ji, Ming, 1994, "Simulations of Atmospheric Variability Induced by Sea Surface Temperatures and Implications for Global Warming," *Science,* Vol. 266, p. 632-634, 28 Oct 1994.

Kurtén, Björn, 1981, *How to Deep-Freeze a Mammoth.*

Laj, Carlo, Guitton, Sylvie, Kissel, Catherine, and Mazaud, Alain, 1988, "Complex behavior of the geomagnetic field during three successive polarity reversals, 11-12 m.y.B.P.," *Journal of Geophys. Res.,* Vol. 93, p. 11,655-11,666, 10 Oct 1988.

Laj, Carlo, Mazaud, Alain, Weeks, Robin, Fuller, Mike, and Herrero-Bervera, Emilio, 1991, "Geomagnetic reversal paths," *Nature,* Vol. 351, p. 447, 6 Jun 1991.

Lamb, H. H., 1972, *Climate, Present, Past, and Future,* Vol. 1, p. 432.

Lamb, H. H., 1982, *Climate, history and the modern world.*

Larson, Roger L. and Pitman, Walter C. III, 1972, "World-wide correlation of Mesozoic Magnetic Anomalies, and its Implications," *Geol. Soc. Am. Bull.,* Vol. 83, p. 3645-3662. (Also in *Megacycles: Long-Term Episodicity in Earth and Planetary History.*)

Lemonick, Michael D., "Rewriting the Book on Dinosaurs," *Time,* p. 42-49, 26 Apr 1993.

Le Pichon, Xavier, 1968, "Marine Magnetic Anomalies, Geomagnetic Field Reversals, and Motions of the Ocean Floor and Continents," *J. Geophys. Res.,* Vol. 73, p. 2119-2136. (Also in *Plate Tectonics and Geomagnetic Reversals,* p. 265-282.)

Le Pichon, Xavier, 1968, "Sea-floor Spreading and Continental Drift," *Journal of Geophys. Res.,* Vol. 73, p. 3661-3697.

Lessem, Don, 1993, "Weird Wonders Fuel a Battle over Evolution's Path," *Smithsonian*, p. 106-115, Jan 1993.

Levi, S., Audusson, H., Duncan, R. A., Kristjansson, L., Gillot, P. Y. and Jakobson, S. P., 1990, "Late Pleistocene geomagnetic excursion in Icelandic lavas: Confirmation of the Laschamp excursion," *Earth and Planet. Sci. Lett.*, Vol. 96, p. 443-457, 1990.

Levi, S. and Karlin, R., 1989, "A sixty thousand year paleomagnetic record from Gulf of California sediments: secular variation, late Quaternary excursions and geomagnetic implications," *Earth and Planet. Sci. Lett.*, Vol. 92, p. 219-233.

Libby, Willard F., 1981, *Talking to People*.

Liddicoat, J. C., 1992, "Mono Lake Excursion in Mono Basin, California, and at Carson Sink and Pyramid Lake, Nevada," *Geophys. J. Int.*, V. 108, p. 442-452.

Liddicoat, J. C. and Coe, R. S., 1979, "Mono Lake geomagnetic excursion," *Journal of Geophys. Res.*, Vol. 84, p. 261-271, 10 Jan 1979.

Lindsay, J. F. and Srnka, L. J., 1975, "Galactic Dust Lanes and Lunar Soil," *Nature*, Vol. 257, p. 776-778, 30 Oct 1975. (Also in the book *Megacycles: Long-Term Episodicity in Earth and Planetary History*.)

Linke, G., Katzenberger, O. and Grün, R., 1985, "Description and ESR Dating of the Holsteinian Interglaciation," *Quaternary Science Rev.*, Vol. 4, p. 319-331.

Lister, Adrian and Bahn, Paul, 1994, *Mammoths*.

Loper, David E. and McCartney, Kevin, 1986, "Mantle Plumes and the Periodicity of Magnetic Field Reversals," *Geophys. Res. Lett.*, Vol. 13, No. 13, p. 1525-1528, Dec 1986.

Loper, David E. and McCartney, Kevin, and Buzyna, George, 1988, "A Model of Correlated Episodicity in Magnetic-Field Reversals, Climate, and Mass Extinctions," *Journal of Geology*, Vol. 96, p. 1-15.

Lowell, Robert P., Roan, Peter A. and Von Herzen, Richard P., 1995, "Seafloor hydrothermal systems," *Journal of Geophys. Res.*, Vol. 100, No. B1, p. 327-352, 10 Jan 1995.

Lowell, T. V., Heusser, C. J., Andersen, B. G., Moreno, P. I., Hauser, A., Heusser, L. E., Schlücter, C., Marchant, D. R. and Denton, G. H., 1995, "Interhemispheric Correlation of Late Pleistocene Glacial Events," *Science*, Vol. 269, p. 1541-1549, 15 Sep 1995.

Lowrie, W., 1989, "Magnetic time scales and reversal frequency," in *Geomagnetism and Palaeomagnetics*," ed. by F. J. Lowes et al., p. 155-183.

Lutz, Richard A. and Haymon, Rachel M., 1994, "Rebirth of a Deep-Sea vent," *National Geographic*, Vol. 186, p. 114-126, Nov 1994.

Lyell, Charles, 1843, "On the upright Fossil Trees found at different levels in the Coal Strata of Cumberland, Nova Scotia," *Am. Journ. of Sci.*, 1:45, p. 353-356.

Lyons, J. B. and Officer, Charles B., 1992, "Mineralogy and petrology of the Haiti Cretaceous/Tertiary section," *Earth Planet. Sci. Lett.*, Vol. 109, p. 205-224.

Macdonald, Ken C., 1989, "Tectonic and magmatic processes on the East Pacific Rise, in Winterer, E. L., Hussong, D. M. and Decker, R. W., eds., *The Eastern Pacific Ocean and Hawaii, the Geology of North America*, Vol. N.

Macdonald, Ken C., Becker, Feir, Spiess, F. N. and Ballard, R. D., 1980, "Hydrothermal heat flux of the 'black smoker' vents on the East Pacific Rise," *Earth and Planet. Sci. Lett.*, Vol. 48, p. 1-7.

Madden, Theodore R., 1979, "Electrical Measurements as Stress-Strain Monitors," *Earthquake Information Bulletin,* Vol. 11, p. 4-8, Jan/Feb 1979.

Malin, S. R. C. and Clark, Anne D., 1974, "Geomagnetic Secular Variation, 1962.5 to 1967.5," *Journ. Royal Astronomical Soc.* Vol. 36, p. 11-20.

Malkus, W. V. R., 1968, "Precession of the Earth as the Cause of Geomagnetism," *Science,* Vol. 160, p. 259-264, 19 Apr 1968.

Mansinka, L. and Smylie, D. E., 1968, "Earthquakes and the Earth's Wobble," *Science,* Vol. 161, p. 1127-1129, 13 Sep 68.

Markson, Ralph and Nelson, Richard, 1970, "Mountain-Peak Potential-Gradient Measurements and the Andes Glow," *Weather,* Vol. 25, p. 350-360, Aug 1970.

Martinson, D. G., Pisias, N. G., Hays, J. D., Imbrie, J., Moore, T. C. Jr. and Shackleton, N., 1987, "Age Dating and the Orbital Theory of the Ice Ages: Development of a High-Resolution 0 to 300,000-year Chronostratigraphy," *Quaternary Research,* Vol. 27, p. 1-29.

Matsuda, Tykihiko, and Uyeda, Seiya, 1971, "On the Pacific-type Orogeny and its Model: Extension of the Paired Belts Concept and Possible Origin of Marginal Seas," *Tectonophysics,* Vol. 11, p. 5-27. (Also in *Plate Tectonics and Geomagnetic Reversals,* p. 632-647.)

Matthews, Robley K., 1969, "Tectonic Implications of Glacio-Eustatic Sea Level Fluctuations," *Earth and Planet. Sci. Lett.* Vol. 5, p. 459-462.

Matthews, Samuel W., 1976, "What's Happening to Our Climate?" *National Geographic,* p. 576-615, Nov 1976.

Matuyama, Motonori, 1929, "On the Direction of Magnetisation of Basalt in Japan, Tyôsen and Manchuria," *Japan Academy Proceedings,* Vol. 5, p. 203-205. (Also in *Plate Tectonics and Geomagnetic Reversals,* p. 154-156.)

Maurrasse, F. and Sen, G., 1991, "Impacts, tsunamis, and the Haitian Cretaceous-Tertiary boundary layer," *Science,* V. 252, p. 1690-1693, 21 Jun 1991.

Maxwell, Arthur E., Von Herzen, Richard P., Hsü, K. Jinghwa, Andrews, James E., Saito, Tsunemasa, Percival, Stephen F. Jr., Milow, E. Dean and Boyce, Robert E., 1970, "Deep Sea Drilling in the South Atlantic," *Science,* Vol. 168, p. 1047-1059. (Also in *Plate Tectonics and Geomagnetic Reversals,* p. 560-582.)

Mayo, L. R. and March, R. S., 1990, "Air temperature and precipitation at Wolverine Glacier, Alaska; glacier growth in a warmer, wetter climate," *Annals of Glaciology,* Vol. 14, p. 191-194.

Mazaud, A., Laj, C., Bard, E. and Arnold, M., 1992, "Geomagnetic Calibration of the Radiocarbon time-scale," in *The Last Deglaciation: Absolute and Radiocarbon Chronologies,* (eds. Bard, E. & Broecker, W. S.) NATO ASI Series 2, 1992.

Mazaud, A., Laj, C., Laurent de Sèze and Verosub, K. L., 1983, "15-Myr periodicity in the frequency of geomagnetic reversals since 100myr," *Nature,* Vol. 304, p. 328-330, 28 Jul 1983.

Mazaud, A., Laj, C., Laurent de Sèze and Verosub, K. L., 1984, Letter to *Nature,* Vol. 311, 27 Sept 1984.

Mazie, David, *National Geographic News Service,* 25 Oct 1994.

McBirney, Alexander R., 1976, "Some Geologic Constraints on Models for Magma Generation in Orogenic Environments," *Can. Mineralogist,* Vol. 14, p. 245-254.

McCrea, W. H., 1975, "Ice ages and the galaxy," *Nature,* Vol. 255, p. 607-609, 19 Jun 1975.

McElhinny, M. W., 1971, "Geomagnetic Reversals during the Phanerozoic," *Science*, Vol. 172, p. 157-159.

McElhinny, M. W. and Senanayaka, W. E., 1982, "Variations in the Geomagnetic Dipole 1: The Past 50,000 years," *J. Geomagn. Geoelec. Kyoto*, V. 34, p. 39-51.

McFadden, Philip L. and Merrill, Ronald T., 1995, "History of Earth's magnetic field and possible connections to core-mantle boundary processes," *Journal of Geophys. Res.*, Vol. 100, No. B1, p. 307-316, 10 Jan 1995.

McFadden, Philip L., Merrill, Ronald T. and McElhinny, M. W., 1988, "Dipole/ Quadrupole Family Modeling of Paleosecular Variation," *Journal of Geophys. Res.*, Vol. 93, p. 11,583-11,588, 10 Oct 1988.

McKenzie, Dan P., 1969, "Speculations on the Consequences and Causes of Plate Motions," *Geophysical Journal of the Royal Astronomical Soc.*, Vol. 18, p. 1-32.

McKenzie, Dan P. and Parker, Robert L., 1967, "The North Pacific: An Example of Tectonics on a Sphere," *Nature*, Vol. 216, p. 1276-1280. (Also in *Plate Tectonics and Geomagnetic Reversals*, p. 57-64.)

McLaren, Digby J., 1982, "Frasnian-Famennian extinctions," *Geol. Soc. Am. Spec. Pap. 190*, p. 477-484.

McLean, Dewey M., 1985, "Deccan Traps Mantle Degassing in the Terminal Cretaceous Marine Extinctions," *Cretaceous Research*, Vol. 6, p. 235-259.

Megacycles: Long-Term Episodicity in Earth and Planetary History, 1981, editor, George E. Williams.

Menard, H. W. and Atwater, Tanya, 1968, "Changes in Direction of Sea Floor Spreading," *Nature*, Vol. 219, p. 463-467, 3 Aug 1968. (Also in *Plate Tectonics and Geomagnetic Reversals*, p. 412-419.)

Merrill, Ronald T. and McElhinny, M. W., 1983, *The Earth's Magnetic Field*.

Merrill, Ronald T. and McFadden, Philip L., 1990, "Paleomagnetism and the Nature of the Geodynamo," *Science*, Vol. 248, p. 345-350, 20 Apr 1990.

Merrill, Ronald T. and McFadden, Philip L., 1995, "Dynamo theory and paleomagnetism," *Jour. Geophys. Res.*, Vol. 100, No. B1, P. 317-326, 10 Jan 1995.

Meyers, Robert A., editor, "Reversals of the Earth's Magnetic Field," *Encyclopedia of Physical Science and Technology*, Vol. 6, p. 118-120.

Miller, Gifford H., 1976, "Anomalous local glacier activity, Baffin Island, Canada: Paleoclimatic implications," *Geology*, Vol. 4, p. 502-504, Aug 1976.

Miller, Gifford H. and de Vernal, Anne, 1992, "Will Greenhouse warming lead to Northern Hemisphere ice-sheet growth?" *Nature*, Vol. 355, p. 244-246, 16 Jan 1992.

Miller, Gifford H., Funder, Svend, de Vernal, Anne, and Andrews, John T., 1992, "Timing and character of the last interglacial-glacial transition in the eastern Canadian Arctic and northwest Greenland," *Geo. Soc. Am. Spec. Pap. 270*, p. 223-231.

Miller, Maynard, "Alaska's mighty rivers of ice," *National Geographic*, Feb 1967.

Mitchell, Jay Murray, 1976, "An overview of climatic variability and its causal mechanisms," *Quat. Res.*, Vol. 6, p. 481-493.

Monastersky, Richard, 1995, "Tropical Trouble: Two decades of Pacific warmth have fired up the globe," *Science News*, Vol. 147, p- 154-155, 11 Mar 1995.

Moore, T. C, Pisias, N. G. and Dunn, D. A., 1982, "Carbonate time series of the Quaternary and late Miocene sediments in the Pacific Ocean: A spectral comparison," *Mar. Geol.*, Vol. 46, p. 217-234.

Moore, T. C., van Andel, Tj. H., Sancetta, C. and Pisias, N., 1978, "Cenozoic Hiatuses in Pelagic Sediments," *Micropaleontology*, Vol. 24, p. 113-138. (Also in *Megacycles: Long-Term Episodicity in Earth and Planetary History*.)

Morell, Virginia, 1993, "How Lethal was the K-T Impact?" *Science*, Vol. 261, p. 1518-1519, 17 Sept 1993.

Morgan, W. Jason, 1968, "Rises, Trenches, Great Faults, and Crustal Blocks," *Journal of Geophys. Res.*, Vol. 73, p. 1959-1982. (Also in *Plate Tectonics and Geomagnetic Reversals*, p. 65-88.)

Mörner, Nils-Axel, 1971, "The Plum Point Interstadial: Age, Climate, and Subdivision," *Can. Journal Earth. Sci.*, Vol. 8, p. 1423-1431.

Mörner, Nils-Axel and Lanser, Johan P., 1974, "Gothenburg magnetic 'flip'," *Nature*, Vol. 251, p. 705-706, 4 Oct 1974.

Mörner, Nils-Axel, Lanser, Johan P., and Hospers, J., 1971, "Late Weichselian Palaeomagnetic Reversal," *Nature Phys. Sci.*, V. 234, p. 173-174, 27 Dec 1971.

Morris, D. and Berge, G. L., 1964, "Direction of the galactic magnetic field in the vicinity of the Sun," *Astrophys. Journal,* Vol. 139, p. 1388-1393.

Murray, Bruce C., Strom, Robert G., Trask, Newell J. and Gault, Donald E., 1975, "Surface History of Mercury: Implications for Terrestrial Planets," *Journal of Geophys. Res.*, Vol. 80, p. 2508-2514, 10 Jun 1975. (Also in the book *Megacycles: Long-Term Episodicity in Earth and Planetary History*.)

Nance, John J., 1989, *On Shaky Ground.*

Neftel, A., Oeschger, H., Stafflebach, T. and Stauffer, B., 1988, "CO_2 record in the Byrd ice core 50,000-5,000 year BP," *Nature*, Vol. 231, p. 609-611, 18 Feb 1988.

Negi, J. G. and Tiwari, R. K., 1983, "Matching Long Term Periodicities of Geomagnetic Reversals and Galactic Motions of the Solar System," *Geophys. Res. Lett.*, Vol. 10, No. 8, p. 713-716, Aug 1983.

Negi, J. G. and Tiwari, R. K., 1984, "Periodicities of palaeomagnetic intensity and palaeoclimatic variations: a Walsh spectral approach," *Earth and Planet. Sci. Lett.*, Vol. 70, p. 139-147, 1984.

Newell, Norman D., 1963, "Crises in the History of Life," *Scientific American*, Feb 1963.

Newell, Norman D., 1967, "Revolutions in the History of Life," *Geol. Soc. Am., Spec. Pap. 89*, p. 63-91.

Newell, Norman D., 1982, "Mass Extinctions - Illusions or Realities?" *Geol. Soc. Am. Spec. Pap. 190*, p. 257-263.

Ninkovitch, Dragoslav, 1968, "Pleistocene volcanic eruptions in New Zealand recorded in deep-sea sediments," *Earth and Planet. Sci. Lett.*, Vol. 4, p. 89-102.

Ninkovitch, Dragoslav, Opdyke, N. D., Heezen, B. C. and Foster, J. H., 1966, "Paleomagnetic stratigraphy, rates of deposition and tephrachronology in North Pacific deep sea sediments," *Earth Planet. Sci. Lett.*, Vol. 1, p. 476-492.

Noyes, Robert W., 1982, *The Sun, Our Star.*

Oberbeck, V. R., Marshall, J. R. and Aggarwal, H., 1993, "Impacts, Tillites, and the Breakup of Gondwanaland," *The Journal of Geology*, Vol. 101, p. 1-19, Jan 1993.

Officer, Charles B., 1990, "Extinctions, iridium and shocked minerals associated with the Cretaceous/Tertiary Transition," *Journal Geological Education*, Vol. 38, p. 402-425.

Officer, Charles B. and Carter, N. L., 1991, "A review of the structure, petrology and dynamic deformation characteristics of some enigmatic terrestrial structures," *Earth-Science Review*, Vol. 30, p. 1-49.

Officer, Charles B. and Drake, Charles L., 1983, "The Cretaceous-Tertiary Transition," *Science*, Vol. 219, p. 1383-1390, 25 Mar 1983.

Officer, Charles B. and Drake, Charles L., 1985, "Epeirogeny on a Short Geological Time Scale," *Tectonics,* Vol. 4, p. 603-612.

Officer, Charles B. and Drake, Charles L., 1985, "Terminal Cretaceous Environmental Events," *Science,* 8 Mar 1985, Vol. 227, p. 1161-1166.

Officer, Charles B. and Drake, Charles L., 1985, answer to Smit, Kyte, and French (in Letters), *Science*, Vol. 230, 13 Dec 1985.

Officer, Charles B., Hallam, Anthony, Drake, Charles L. and Devine, Joseph D., 1987, "Late Cretaceous and paroxysmal Cretaceous/Tertiary extinctions," *Nature* Vol. 326, p. 143-149, 12 Mar 1987.

Oltmans, S. J. and Hoffman, D. J., 1995, "Increase in lower-stratospheric water vapour at a mid-latitude Northern Hemisphere site from 1981 to 1994," *Nature,* Vol. 374, p. 146-149, 9 Mar 1995.

Opdyke, N. D., Glass, B., Hays, J. D. and Foster, J., 1966, "Paleomagnetic Study of Antarctic Deep-Sea Cores," *Science*, Vol. 154, p. 349-357, 21 Oct 1966.

Öpik, E. J., 1970, "The Ice Ages," in *Adventures in Earth History*, ed. by Preston Cloud, p. 870-877.

Öpik, E. J., 1976, "Solar Structure, Variability, and the Ice Ages (Solar Variability and Climate)," *Astron. Journal* Vol. 12, p. 253-276. (Also in the book *Megacycles: Long-Term Episodicity in Earth and Planetary History*.)

Oppenheimer, D. et al., 1993, "The Cape Mendocino, California, Earthquakes of April 1992: Subduction at the Triple Junction," *Science*, Vol. 261, p. 443-438, 23 Jul 1993.

Owen, Robert M. and Rea, David K., 1985, "Sea-floor hydrothermal activity links climate to tectonics: The Eocene carbon dioxide greenhouse," *Science*, Vol. 227, p. 166-169, 11 Jan 1985.

Pakiser, L. C., Eaton, J. P., Healy, J. H. and Raleigh, C. B., 1969, "Earthquake Prediction and Control," *Science*, Vol. 166, p. 1467-1474, 19 Dec 1969.

Pal, P. C. and Creer, Kenneth M., 1986, "Geomagnetic reversal spurts and episodes of extraterrestrial catastrophism," *Nature*, Vol. 320, p. 148-150, 13 Mar 1986.

Pardee, J. T., 1942, "Unusual currents in glacial Lake Missoula, Montana," *Geol. Soc. Am. Bull,* Vol. 53, p. 1569-1600.

Paterson, W. S. B., Koerner, R. M., Fisher, D., Johnsen, S. J., Clausen, H. B., Dansgaard, W., Bucher, P. and Oeschger, H., 1977, "An oxygen-isotope climatic record from the Devon Island ice cap, arctic Canada," *Nature*, Vol. 266, p. 508-511, 7 Apr 1977.

Peteet, D., Rind, D. and Kukla, George J., 1992, "Wisconsin ice-sheet initiation: Milankovitch forcing, paleoclimatic data, and global climate modeling," *Geol. Soc. Am. Special Pap. 270*, p. 53-69.

Peterman, Zell E., Hedge, Carl E. and Tourtelot, Harry A., 1970, "Isotopic composition of strontium in sea water throughout Phanerozoic time," *Geochim. et Cosmochim. Acta* Vol. 34, p. 105-108 and 111-118. (Also in *Megacycles: Long-Term Episodicity in Earth and Planetary History*.)

Pillmore, C. L., Tschudy, R. H., Orth, C. J., Gilmore, J. S. and Knight, J. D., 1984, "Geologic Framework of Nonmarine Cretaceous-Tertiary Boundary Sites, Raton Basin, New Mexico and Colorado," *Science*, Vol. 223, p. 1180-1183, 16 Mar 1984.

Piper, J. D. A., 1974, "Proterozoic Crustal Distribution, mobil belts and apparent polar movements," *Nature*, Vol. 251, p. 381-384.

Pitman, Walter C. III, and Hayes, Dennis E., 1968, "Sea-floor Spreading in the Gulf of Alaska," *Journal of Geophys. Res.*, Vol. 73, p. 6571-6580. (Also in *Plate Tectonics and Geomagnetic Reversals*, p. 420-429.)

Plafker, George, 1965, "Tectonic Deformation Associated with the 1964 Alaska Earthquake," *Science*, Vol. 148, p. 1675-1687. (Also in *Plate Tectonics and Geomagnetic Reversals*, p. 311-331.)

Pollard, D., 1978, "An investigation of the astronomical theory of the ice ages using a simple climate-ice sheet model," *Nature*, Vol. 272, p. 233-235.

Prévot, Michel, Mankinen, E. A., Coe, R. S. and Grommé, C. S., 1985, "The Steens Mountain (Oregon) Geomagnetic Polarity Transition 2. Field Intensity Variations and Discussions of Reversal Models," *Journal of Geophys. Res.*, Vol. 90, p. 10,417-10,448, 10 Oct 1985.

Pszczolkowski, A., 1986, *Bull. Polish Acad. Sci. Earth Sci.*, Vol. 34, no. 1, p. 81.

Raisbeck, G. M., Yiou, F., Bourles, D., Lorius, C., Jouzel, J. and Barkov, N. I., 1987, "Evidence for two intervals of enhanced [10]Be deposition in Antarctic ice during the last glacial period," *Nature*, Vol. 326, p. 273-277, 19 Mar 1987.

Raisbeck, G. M., Yiou, F., Fruneau, M., Loiseaux, J. M., Lieuvin, M., Ravel, J. C. and Hays, J. D., 1979, "A search in a marine sediment core for [10]Be concentration variations during a geomagnetic field reversal," *Geophys. Res. Lett.*, p. 717-719, Sep 1979.

Rampino, Michael R., 1979, "Possible relationships between changes in global ice volume, geomagnetic excursions, and the eccentricity of the Earth's orbit," *Geology*, Vol. 7, p. 584-587, Dec 1979.

Rampino, Michael R., 1981, "Revised age estimates of Brunhes palaeomagnetic events: Support of a link between geomagnetism and orbital eccentricity variations," *Geophys. Res. Lett.*, Vol. 8, p. 1047-1050.

Rampino, Michael R. and Reynolds, Robert C., 1983, "Clay Mineralogy of the Cretaceous-Tertiary Boundary Clay," *Science*, Vol. 219, p. 495-498, 4 Feb 1983.

Rampino, Michael R. and Stothers, Richard B., 1984, "Geological Rhythms and Cometary Impacts," *Science*, Vol. 226, p. 1427-1431, 21 Dec 1984.

Rampino, Michael R. and Stothers, Richard B., 1984, "Terrestrial mass extinctions, cometary impacts and the Sun's motion perpendicular to the galactic plane," *Nature*, Vol. 308, p. 709-712, 19 Apr 1984.

Rampino, Michael R. and Stothers, Richard B., 1988, "Flood Basalt Volcanism During the past 250 Million Years," *Science*, Vol. 241, p. 663-668, 5 Aug 1988.

Rankama, K., 1954, "The isotopic constitution of carbon in ancient rocks as an indicator of its biogenic or non-biogenic origin," *Geochim. Cosmochim. Acta*, Vol. 5, p. 142-152.

Raup, David M., 1985, "Magnetic Reversals and mass extinctions," *Nature*, Vol. 314, p. 341-343, 28 Mar 1985.

Raup, David M., 1986, *The Nemesis Affair*.

Raup, David M., 1991, *Extinction: Bad Genes or Bad Luck?*

Raup, David M. and Jablonski, D., 1993, "Geography of End-Cretaceous Marine Bivalve Extinctions," *Science*, Vol. 260, p. 971-973, 14 May 1993.

Raup, David M. and Sepkoski, J. John Jr., 1984, "Periodicity of extinctions in the geologic past," *Proc. Natl. Acad. Sci.*, Vol. 81, p. 801-805, Feb 1984.

Raup, David M. and Stanley, Steven M., 1978, *Principles of Paleontology*.

Ray, Dixy Lee with Guzzo, Louis R., 1990, *Trashing the Planet*.

Ray, Dixy Lee with Guzzo, Louis R., 1993, *Environmental Overkill, Whatever Happened to Common Sense?*

Raynaud, D., Chappellaz, J., Barnola, J. M., Korotkevich, Y. S. and Lorius, C., 1988, "Climatic and CH_4 cycle implications of glacial-interglacial CH_4 change in the Vostok ice core," *Nature*, Vol. 333, p. 655-677, 16 Jun 1988.

Raynaud, D., Jouzel, J., Barnola, J. M., Chappellaz, J., Delmas, R. J. and Lorius, C., 1993, "The Ice Record of Greenhouse Gases," *Science*, Vol. 259, p. 926-934, 12 Feb 1993.

Rea, D. K. and Vallier, T. L., 1983, "Two Cretaceous volcanic episodes in the western Pacific Ocean," *Geol. So. Am. Bull.*, Vol. 94. p. 1430-1437, Dec 1983.

Regan, Robert D. and Hinze, William J., 1975, "Gravity and Magnetic Investigations of Meteor Crater, Arizona," *Journal of Geophys. Res.*, Vol. 80, p. 776-788, 10 Feb 1975.

Reid, G. C., Isaksen, I. S. A., Holzer, T. E. and Crutzen, P. J., 1976, "Influence of ancient solar-proton events on the evolution of life," *Nature*, Vol. 259, p. 177-179, 22 Jan 1976.

Renne, Paul R. and Basu, Asish R., 1991, "Rapid Eruption of the Siberian Traps Flood Basalts at the Permo-Triassic Boundary," *Science*, Vol. 253, p. 176-179, Jul 1991.

Reynolds, Richard L., 1977, "Paleomagnetism of Welded Tuffs of the Yellowstone Group," *Journal of Geophys. Res.*, Vol. 82, p. 3677-3693, 10 Sep 1977.

Rich, J. E., Johnson, G. L., Jones, J. E. and Campsie, J., 1986, *Paleoceanography*, Vol. 1, p. 85.

Rikitake, Tsuneji, 1968, "Earthquake Prediction," *Earth-Sci. Rev.*, Vol. 4, p. 245-282.

Rikitake, Tsuneji, 1971, "Electric Conductivity Anomaly in the Earth's Crust and Mantle," *Earth-Science Rev.*, Vol. 7, p. 35-65.

Ringrose, P.S., 1989, "Palaeoseismic (?) liquefaction event in late Quaternary lake sediment at Glen Roy, Scotland," *Terra Nova*, Vol. 1, p. 57-62, 1989.

Robin, E., Frogel, L., Jéhanno, C. and Rocchia, R., 1993, "Evidence for a K/T Impact event in the Pacific Ocean," *Nature*, Vol. 363, p. 615-617, 17 Jun 1993.

Roemmich, Dean and McGowan, John, 1995, "Climatic Warming and the Decline of Zooplankton in the California Current," *Science*, Vol. 267, p. 1324-1326, 3 Mar 1995.

Rona, Peter A., 1973, "Relations between rates of sediment accumulation on continental shelves, sea-floor spreading and eustacy inferred from the central North Atlantic," *Geol. Soc. Am. Bull.*, Vol. 84, p. 2851-2872, 1 Sep 1973.

Roperch, P., Bonhommet, N. and Levi, S., 1988, "Paleointensity of the earth's magnetic field during the Laschamp excursion and its geomagnetic implications," *Earth and Planet. Sci. Lett.*, Vol. 88, p. 209-219, 1988.

Rubinshteyn, M. M., 1967, "Orogenic phases and the periodicity of folding in the light of absolute age measurements," *Geotectonics*, p. 80-85.

Ruddiman, W. F. and McIntyre, A., 1979, "Warmth of the Subpolar North Atlantic Ocean During Northern Hemisphere Ice-Sheet Growth," *Science*, Vol. 204, p. 173-175, 13 Apr 1979.

Ruddiman, W. F. and McIntyre, A., 1981, "Oceanic Mechanisms for Amplification of the 23,000-Year Ice-Volume Cycle, *Science*, Vol. 212, p. 617-627, 8 May 1981.

Ruddiman, W. F., McIntyre, A., Niebler-Hunt, V. and Durazzi, J. T., 1980, "Oceanic Evidence for the Mechanism of Rapid Northern Hemisphere Glaciation," *Quaternary Research*, Vol. 13, p. 33-64.

Ruddiman, W. F. and McIntyre, A., 1981, "The North Atlantic Ocean during the last Deglaciation," *Palaeogeography, Palaeoclimatology, Palaeoecology*, Vol. 35, p. 145-214.

Ruddiman, W. F. and McIntyre, A., 1982, "Severity and Speed of Northern Hemisphere glaciation pulses: The limiting case?" *Bull. Geol. Soc. Am.*, Vol. 93, p. 1273.

Runcorn, S. K., 1962, "Convection Currents in the Earth's Mantle," *Nature*, Vol. 195, p. 1248-1249. (Also in *Megacycles: Long-Term Episodicity in Earth and Planetary History*.)

Runcorn, S. K., 1962, "Towards a Theory of Continental Drift," *Nature*, Vol. 193, p. 311-314, 27 Jan 1962.

Runcorn, S. K., Collinson, D. W., O'Reilly, W., Stephenson, A., Greenwood, N. N. and Battey, M. H., 1970, "Magnetic Properties of Lunar Samples," *Science*, Vol. 167, p. 697-700, 30 Jan 1970.

Runcorn, S. K. and Urey, H. C., 1973, "Abstract-concerns remanent magnetism on the moon," *Science*, Vol. 180, p. 636-638.

Runcorn, S. K. and Urey, H. C., 1973, "A New Theory of Lunar Magnetism," *Science*, Vol. 180, p. 636-638, May 1973.

Ryan, W. B. F. and Cita, M. B., 1977, "Ignorance concerning episodes of ocean-wide stagnation," *Mar. Geol.*, Vol. 23, p. 197-215.

Sagan, Carl, Toon, O. and Gierash, P., 1973, "Climatic change on Mars," *Science*, Vol. 181, p. 1045-1049.

Schlanger, S. O., Jenkyns, H. C. and Premoli-Silva, I., 1981, "Volcanism and vertical tectonics in the Pacific Basin related to global Cretaceous transgressions," *Earth Planet. Sci. Lett.*, Vol. 52, p. 435-449.

Schneider, Eric D. and Vogt, Peter R., 1968, "Discontinuities in the History of Sea-Floor Spreading," *Nature*, Vol. 217, p. 1212-1222.

Schnepp, Elisabeth and Hradetzky, Helmuth, 1994, "Combined paleointensity and $^{40}Ar/^{39}Ar$ age spectrum data from volcanic rocks of the West Eifel (Germany): Evidence for an early Brunhes geomagnetic excursion," *Journal of Geophys. Res.*, Vol. 99, p. 9061-9076, 10 May 1994.

Schwartz, Richard D. and James, Philip B., 1984, "Periodic mass extinctions and the Sun's oscillation about the galactic plane," *Nature*, Vol. 308, p. 712-713, 19 Apr 1984.

Sclater, John G. and Tapscott, Christopher, 1979, "The History of the Atlantic," *Scientific American*, Vol. 240, No. 6, June 1979, p. 156-174.

Sepkoski, J. John Jr., 1982, "Mass Extinctions in the Phanerozoic oceans: A review," *Geol. Soc. Am. Spec. Pap. 190*, p. 283-289.

Shackleton, Nicholas J., 1982, "The Deep-sea Sediment record of Climate Variability, *Progr. in Oceanogr.* Vol. 11, p. 199-218.

Shackleton, Nicholas J. and Opdyke, N. D., 1977, "Oxygen isotope and paleomagnetic evidence for early northern hemisphere glaciation," *Nature*, Vol. 270, p. 216-219.

Sharpton, Virgil L., Burke, K., Camargo-Zanoguera, A., Hall, S. A., Lee, D. S., Marín, L. E., Suárez-Reynoso, G., Quezada-Muñeton, J. M., Spudis, P. D. and Urratia-Fucugauchi, Jaime, 1993, "Chicxulub Multiring Impact Basin: Size and other Characteristics Derived from Gravity Analysis," *Science*, Vol. 261, p 1564-1567, 17 Sep 1993.

Shaw, Evelyn, 1977, "Can Animals Anticipate Earthquakes?" *Natural History*, Vol. 86, No. 9, p. 14-20, Nov 1977.

Shaw, H. R., Kistle, R. W. and Evernden, J. F., 1971, "Sierra Nevada Plutonic Cycle: Part II: Tidal Energy and a Hypothesis for Orogenic-Epeirogenic Periodicities," *Geol. Soc. Am. Bull.*, Vol. 82, p. 869-896, April 1971.

Shaw, H. R. and Moore, J. G., 1988, "Magmatic Heat and the El Niño cycle," *Eos*, Vol. 69, p. 1553-1565, 8 Nov 1988.

Sheridan, Robert E., 1983, "Phenomena of pulsation tectonics related to the break-up of the eastern North American continental margins," *Tectonophysics*, Vol. 94, p. 169-185.

Sheridan, Robert E., 1986, "Pulsation tectonics as the control of North Atlantic palaeoceanography," in *North Atlantic Paleoceanography, Geological Soc. Spec. Publ. No. 21*, editors Summerhayes and Shackleton, p. 255-275.

Silverberg, Robert, 1970, *Mammoths, Mastodons and Man.*

Simpson, John F., 1966, "Short Notes, Evolutionary Pulsations and Geomagnetic Polarity," *Geol. Soc. Am. Bull.*, Vol. 77, p. 197-204, Feb 1966.

Sloss, L. L., 1976, "Areas and Volumes of Cratonic Sediments, Western North America and Eastern Europe," *Geology*, Vol. 4, p. 272-276. (Also in *Megacycles: Long-Term Episodicity in Earth and Planetary History.*)

Sloss, L. L. and Speed, Robert C., "Relationships of Cratonic and Continental-Margin Tectonic Episodes," from "Tectonics and Sedimentation," ed. W. R. Dickinson, *Soc. Econ. Paleont. Mineral Spec. Publ. No. 22*, p. 98-119.

Smit, Jan, 1982, "Extinction and evolution of planktonic foraminifera after a major impact at the Cretaceous/Tertiary boundary," *Geol. Soc. Am. Spec. Pap. 190*, p. 329-352.

Smit, Jan, Montanari, A., Swinburne, N. H. M., Alvarez, W., Hildebrand, A., Margolis, S., Claeys, P., Lowrie, W., Asaro, F., 1992, "Tektite-bearing, deep-water clastic unit at the Cretaceous-Tertiary boundary in northeastern Mexico," *Geology*, Vol. 20, p. 99-103, Feb 1992.

Smit, Jan and van der Kaars, S., 1984, "Terminal Cretaceous extinctions in the Hell Creek area: compatible with catastrophic extinctions," *Science*, Vol. 223, p. 1177-1179, 16 Mar 1984.

Smith, B. E. and Johnston, M. J. S., 1976, "A tectonomagnetic effect observed before a 5.2 earthquake near Hollister, California," *Journal of Geophys. Res.*, Vol. 81, p. 3556-3560.

Smith, Peter J., editor, 1986, *The Earth.*

Stanley, Steven M., 1984, "Mass Extinctions in the Ocean," *Scientific American,* p. 64-72, June 1984.

Stanley, Steven M., 1984, "Temperature and biotic crises in the marine realm," *Geology*, Vol. 12, p. 205-208, Apr 1984.

Stanley, Steven M., 1987, *Extinction.*

Stanley, Steven M., 1988, "Paleozoic Mass Extinctions: Shared Patterns Suggest Global Cooling as a Common Cause," *Am. Journal of Sci.*, Vol. 288, p. 334-352.

Stauffer, B., Lochbronner, E., Oeschger, H. and Schwander, J., 1988, "Methane concentration in the glacial atmosphere was only half that of the preindustrial Holocene," *Nature*, Vol. 332, p. 812-814, 28 Apr 1988.

Steiner, Johann, 1973, "Possible galactic causes for synchronous sedimentation sequences of the North American and eastern European cratons," *Geology*, Vol. 1, p. 89-92.

Steiner, Johann, 1977, "An expanding Earth on the basis of sea-floor spreading and subduction rates," *Geology*, Vol. 5, p. 313-318, May 1977.

Steiner, Johann, 1978, "Lead isotope events of the Canadian shield, *ad hoc* solar galactic orbits and glaciations," *Precambrian Research*, Vol. 6, p. 269-274.

Steiner, Johann, and Grillmair, E., "Possible Galactic Causes for Periodic and Episodic Glaciations," *Geol. Soc. Am. Bull.*, Vol. 84, Mar 1973, p. 1003-1018.

Stewart, John Massey, "Frozen Mammoths from Siberia Bring Ice Ages to Vivid Life," *Smithsonian*, p. 60-69, Dec 1977.

Stock, Christina, 1995, "Ancient, erratic changes in sea level suggest a coming swell," *Scientific American*, p. 21-22, Aug 1995.

Stockwell, C. H., 1968, "Geochronology of stratified rocks of the Canadian Shield," *Canadian Journal of Earth Sci.* Vol. 5, p. 693-698.

Stommel, Henry and Stommel, Elizabeth, 1979, "The Year without a Summer," *Scientific American*, p. 176-186, Jun 1979.

Stothers, Richard B., 1986, "Periodicity of the Earth's magnetic reversals," *Nature,* Vol. 322, p. 444-446, 31 Jul 1986.

Street-Perrott, F. Alayne and Perrott, R. Alan, 1990, "Abrupt Climate fluctuations in the tropics: the influence of Atlantic Ocean circulation," *Nature*, Vol. 343, p. 607-612, 15 Feb 1990.

Stuiver, Minze, 1971, "Evidence for the variation of atmospheric C^{14} content in the Late Quaternary," *Late Cenozoic Glacial Ages*, Karl K. Turekian, editor.

Suarez, M. J. and Held, I. M., 1976, "Modelling climatic response to orbital parameter variations," *Nature*, Vol. 263, p. 46-47, 2 Sep 1976.

Suess, Hans E., 1965, "Secular Variations of the Cosmic-Ray-Produced Carbon 14 In the Atmosphere and Their Interpretations," *Journal of Geophys. Res.*, Vol. 70, No. 23, p. 5937-5952, 1 Dec 1965.

Sutton, J., 1963, "Long-term Cycles in the evolution of Continents," *Nature*, Vol. 198, p. 731-735.

Sutton, J., 1976, Introductory remarks in "A discussion on Global Tectonics in Proterozoic Times," J. Sutton, R. M. Shackleton, and J. C. Briden, organizers, *Royal Soc. London Philos. Trans. Ser. A 280*, p. 399-403.

Swinbanks, David, 1995, "Japan jumps on board the VAN wagon," *Nature*, Vol. 375, p. 617, 22 Jun 1995.

Swisher, Carl C. et al., 1992, "Coeval ^{40}Ar/^{39}Ar Ages of 65.0 Million Years Ago from Chicxulub Crater Melt Rock and Cretaceous-Tertiary Boundary Tektites," *Science*, Vol. 257, p. 954-958, 14 Aug 1992.

Sykes, Lynn R., 1967, "Mechanism of Earthquakes and Nature of Faulting on The Mid-ocean Ridges," *Journal of Geophys. Res.*, Vol. 72, p. 2131-2153. (Also in *Plate Tectonics and Geomagnetic Reversals*, p. 332-357.)

Taylor, K. C., Lamorey, G. W., Doyle, G. A., Alley, R. B., Grootes, P. M., Mayewski, P. A., White, J. W. C. and Barlow, L. K., 1993, "The 'flickering switch' of late Pleistocene climate changes," *Nature*, Vol. 361, p. 432-436, 4 Feb 1993.

Tazief, Haroun, 1992, *Earthquake Prediction*.

"The Wobbling Earth," *Science News*, Vol. 100, p. 108, 4 Aug 1971.

Thierstein, Hans R. and Berger, Wolfgang, H., 1978, "Injection events in ocean history," *Nature*, Vol. 276, p. 461-466, 30 Nov 1978.

Thomas, Rob, 1993, Personal Communication.

Thompson, G. A., 1960, "Problems of late Cenozoic structure of the basin ranges," *21st Int. Geol. Congr., Copenhagen*, Vol. 18, p. 62-68.

Thorarinsson, Sigurdur, 1965, "Surtsey: Island Born of Fire," *National Geographic*, p. 713-726, May 1965.

Tollman, Edith Kristen-, and Tollman, Alexander, 1993, *Und die Sinflut gab es doch* [German].

Tollman, Edith Kristen-, and Tollman, Alexander, 1994, "The youngest big impact of Earth deduced from geological and historical evidence, *Terra Nova*, Vol. 6, p. 209-217. (English summary of above.)

Tolmachoff, I. P., 1929, "The Carcasses of the Mammoth and Rhinoceros found in the frozen ground of Siberia," *American Philosophical Society Trans.*, p. 1-74.

Tributsch, Helmut, 1978, "Do Aerosol Anomalies Precede Earthquakes?" *Nature*, Vol. 276, p. 606-608, 7 Dec 1978.

Tributsch, Helmut, 1982, "Seismic Sense: Something in the Air Panics Animals Before an Earthquake," *The Sciences*, Vol. 22, p. 24-28, Dec 1982.

Tschudy, R. H., Pellmore, C. L., Orth, C. J., Gilmore, J. S. and Knight, J. D., 1984, "Disruption of the Terrestrial Plant Ecosystem at the Cretaceous-Tertiary Boundary, Western Interior," *Science*, Vol. 225, p. 1030-1032.

Uffen, Robert J., 1963, "Influence of the Earth's Core on the Origin and Evolution of Life," *Nature*, Vol. 198, p. 143-144, 13 Apr 1963.

Umbgrove, J. H. F., 1939, "On Rhythms in the History of the Earth," *Geological Magazine*, Vol. 76, p. 116-129, 1939.

Umbgrove, J. H. F., 1940, "Periodicity in terrestrial processes," *Am. Journal of Sci.*, Vol. 238, p. 573-576, 1940.

Umbgrove, J. H. F., 1950, "Rhythm and synchronism of tectonic movement," *Am. Journal of Sci.*, Vol. 248, p. 521-526, Aug 1950.

Urey, Harold C., 1952, *The Planets*.

Vail, P. R., Mitchum, R. M. Jr. and Thompson, S. III, 1977, "Seismic Stratigraphy and Global Changes of Sea Level, Part 4: Global Cycles of Relative changes of Sea Level," *Am. Assoc. Petroleum Geologists Mem. 26*, p. 83-97. (Also in *Megacycles: Long-Term Episodicity in Earth and Planetary History*.)

Valet, J. P., Laj, C. and Langereis, C. G., 1988, "Sequential geomagnetic reversals recorded in upper Tortonian marine clays in western Crete (Greece)," *Journal of Geophys. Res.* Vol. 93, p. 1131-1151, 1988.

Van, Jon, "Antarctic Ice Grows, Expert Says," the *Chicago Tribune*, 29 May 1986.

Varotsos, P. and Alexopoulous, K., 1984, "Physical Properties of the Variations of the Electric Field of the Earth Preceding Earthquakes, I," *Tectonophysics*, Vol. 110, p. 73-98.

Varotsos, P. and Alexopoulous, K., 1984, "Physical Properties of the Variations of the Electric Field of the Earth Preceding Earthquakes, II. Determination of epicenter and magnitude," *Tectonophysics,* Vol. 110, p. 99-125.

Varotsos, P., Alexopoulous, K., Nomicos, K. and Lazaridou, M., 1986, "Earthquake prediction and electric signals," *Nature*, Vol. 322, p. 120, 10 Jul 1986.

Varotsos, P., Alexopoulous, K., Nomicos, K. and Lazaridou, M., 1988, "Official earthquake prediction procedure in Greece," *Tectonophysics*, Vol. 152, p. 193-196.

Velikovsky, Immanuel, 1950, *Worlds in Collision*.

Velikovsky, Immanuel, 1955, *Earth in Upheaval*.

Verosub, K. L. and Banarjes, S. K., 1977, "Geomagnetic excursions and their paleomagnetic record," *Reviews of Geophysics and Space Physics*, Vol. 15, p. 145-155.

Vine, Fred J., 1966, "Spreading of the Ocean Floor: New Evidence," *Science*, Vol. 154, p. 1405-1415, 16 Dec 1966.

Vine, Fred J. and Matthews, Drummond H., 1963, "Magnetic Anomalies over Oceanic Ridges," *Nature*, Vol. 199, p. 947-949. (Also in *Plate Tectonics and Geomagnetic Reversals*, p. 232-237.)

Vine, Fred J. and Wilson, J. Tuzo, 1965, "Magnetic Anomalies over A Young Oceanic Ridge off Vancouver Island," *Science*, Vol. 150, p. 485-489. (Also in *Plate Tectonics and Geomagnetic Reversals*, p. 238-244.)

Vogt, Peter R., 1972, "Evidence for Global Synchronism in Mantle Plume Convection, and Possible Significance for Geology," *Nature,* Vol. 240, p. 338-342, 8 Dec 1972.

Vogt, Peter R., 1975, "Changes in Geomagnetic Reversal Frequency at times of Tectonic Change: Evidence for Coupling between core and upper mantle processes," *Earth and Planet. Sci. Lett.,* Vol. 25, p. 313-321.

Vogt, Peter R., 1979, "Global magmatic episodes: new evidence and implications for the steady-state mid-oceanic ridge," *Geology*, Vol. 7, p. 93-98, Feb 1979.

Vogt, Peter R., Avery, Otis E., Schneider, Eric D., Anderson, Charles N. and Bracey, Dewey R., 1969, "Discontinuities in Sea-Floor Spreading," *Tectonophysics*, Vol. 8, p. 285-317.

Waddington, C. J., 1967, "Paleomagnetic field reversals and cosmic radiation," *Science*, Vol. 158, p. 913-915.

Walker, Bryce, 1980, *Earthquake,* Time-Life Books.

Walker, Daniel A., 1988, "Seismicity of the East Pacific Rise: Correlations With the Southern Oscillation Index?," *Eos*, Vol. 69, p. 857, 865-867, 20 Sep 1988.

Walker, Daniel A., 1995, "More Evidence Indicates Link Between El Niños and Seismicity," *Eos*, p. 33,34 & 36, 24 Jan 1995.

Ward, Peter D., 1992, *On Methuselah's Trail.*

Ware, B. R., 1992, "Electrophoresis," *Encyclopedia of Science and Technology.*

Warren, G., 1962, *Geol. Surv. N.Z.* Bull. 71, p. 117.

Warwick, James, 1982, "Radio Signals Before Quakes?" *Science News*, Vol. 121, p. 200, 20 Mar 1982.

Watkins, N. D., 1965, "Frequency of extrusion of some Miocene lavas in Oregon during an apparent transition of the polarity of the geomagnetic field," *Nature*, Vol. 206, p. 801-803, 22 May 1965.

Watkins, N. D. and Goodell, H. G., 1967, "Geomagnetic Polarity Change and Faunal Extinction in the Southern Ocean," *Science*, Vol. 156, p. 1083-1087, 26 May 1967.

Weertman, J., 1976, "Milankovitch solar radiation variations and ice age ice sheet sizes," *Nature*, Vol. 26, p. 17-20, 6 May 1976.

Weissel, Jeffrey K. and Hayes, Dennis E., 1971, "Asymmetric Seafloor Spreading South of Australia," *Nature*, Vol. 231, p. 518-521. (Also in *Plate Tectonics and Geomagnetic Reversals*, p. 430-438.)

Weyl, Peter K., 1972, "The Salinity of the North Atlantic Ocean and the Next Glaciation," *Quaternary Research*, Vol. 2, p. 399-400.

Wezel, F. C., Vannucci, S. and Vannucci, R., 1981, "Decouverte de divers niveaux riches en iridium dans la "Scaglia Rossa" et la "Scaglis bianca" de l'Appennin d'Ombrie-Marches (Italie)," *C.R. Acad. Sci. Paris*, Sér. II, 293, p. 837-844.

Whipple, Fred. L., 1964, "The History of the Solar System," in *Adventures in Earth History*, edited by Preston Cloud, p. 100-120, 1970. Also in *Proc. Nat. Acad. Sci.*, Vol. 52, p. 565-594, 1964.

Whitmire, D. P. and Jackson, A. A. IV., 1984, "Are periodic mass extinctions driven by a distant solar companion?" *Nature*, V. 308, p. 713-715, 19 Apr 1984.

Whyte, Martin A., 1977, "Turning Points in Phanerozoic History," *Nature*, Vol. 267. p. 679-682. (Also in *Megacycles: Long-Term Episodicity in Earth and Planetary History*.)

Williams, George E., 1972, "Geological Evidence relating to the origin and secular rotation of the solar system," *Modern Geol.*, Vol. 3, p. 165-173 and 178-181. (Also in *Megacycles: Long-Term Episodicity in Earth and Planetary History*.)

Williams, George E., 1975, "Possible Relation between Periodic Glaciation and the Flexure of the Galaxy," *Earth and Planet. Sci. Lett.*, Vol. 26, p. 361-369. (Also in *Megacycles: Long-Term Episodicity in Earth and Planetary History*.)

Williams, George E., 1981, editor, *Megacycles: Long-Term Episodicity in Earth and Planetary History*.

Wilson, J. Tuzo, 1965, "A New Class of Faults and their Bearing on Continental Drift," *Nature*, Vol. 207, p. 343-347.

Wilson, J. Tuzo, 1966, "Did the Atlantic close and then re-open?" *Nature*, Vol. 211, p. 676-681. (Also in *Megacycles: Long-Term Episodicity of Earth and Planetary History*.)

Winograd, I. J., Coplen, T. B., Landwehr, J. M., Riggs, A. C., Ludwig, K. R., Szabo, B. J., Kolesar, P. T. and Revesz, K. M., 1992, "Continuous 500,000-year Climate Record from Vein Calcite in Devils Hole, Nevada," *Science*, Vol. 258, p. 255-260, 9 Oct 1992.

Woillard, Geneviève M., 1978, "Grande Pile Peat Bog: A Continuous Pollen Record for the Last 140,000 Years," *Quaternary Research*, Vol. 9, p. 1-21.

Woillard, Geneviève M., 1979, "Abrupt end of the last interglacial s.s. in northeast France," *Nature*, Vol. 281, p. 558-562, 18 Oct 1979.

Woillard, Geneviève M. and Mook, Willem G., 1982, "Carbon-14 Dates at Grande Pile: Correlation of Land and Sea Chronologies," *Science,* Vol. 215, p. 159-161, 8 Jan 1982.

Wolbach, Wendy S., Lewis, Roy S. and Anders, Edward, 1985, "Cretaceous Extinctions: Evidence for Wildfires and Search for Meteoritic Material," *Science,* Vol. 230, p. 167-170, 11 Oct 1985.

Wollin, Goesta, Ericson, D. B. and Ryan, W. B. F., 1971, "Variations in Magnetic Intensity and Climatic Changes," *Nature*, Vol. 232, 549-551, 20 Aug 1971.

Wollin, Goesta, Ericson, D. B., Ryan, W. B. F., and Foster, J. H., 1971, "Magnetism of the Earth and Climatic Changes," *Earth and Planet Sci. Lett.*, Vol. 12, p. 175-183.

Wollin, Goesta, Kukla, G. J., Ericson, D. B., Ryan, W. B. F. and Wollin, J., 1973, "Magnetic Intensity and Climatic Changes 1925-1970," *Nature*, Vol. 242, p. 34-37.

Wollin, Goesta, Ryan, W. B. F. and Ericson, D. B., 1977, "Paleoclimate, paleomagnetism and the eccentricity of the Earth's orbit," *Geophys. Res. Lett.*, Vol. 4, p. 267-270, July 1977.

Wollin, Goesta, Ryan, W. B. F. and Ericson, D. B., 1978, "Climate changes, magnetic intensity variations and fluctuations of the eccentricity of the Earth's orbit during the past 2,000,000 years and a mechanism which may be responsible for the relationship," *Earth and Planet. Sci. Lett.*, Vol. 41, p. 395-397.

Wright, G. F., 1900, *The Ice Age in North America*, p. 524.

Yamazaki, Toshitsugu and Ioka, Noboru, 1994, "Long-term secular variation of the geomagnetic field during the last 200 kyr recorded in sediment cores from the western equatorial Pacific," *Earth Planet. Sci. Lett.,* Vol. 128, p. 527-544, Dec 1994.

Zwally, H. Jay, 1989, "Growth of Greenland Ice Sheet: Interpretation," *Science*, Vol. 246, p. 1589-1591, 22 Dec 1989.

Zwally, H. Jay, 1989, "New Information on Changes in the Greenland Ice Sheet," *Eos*, p. 1002, 24 Oct 1989.

Zwally, H. Jay, Brenner, Anita C., Major, Judy A., Bindschadler, Robert A. and Marsh, James, G., 1989, "Growth of Greenland Ice Sheet: Measurement," *Science*, Vol. 246, p. 1587-1589, 22 Dec 1989.

INDEX

YUEN LUI

ROBERT W. FELIX is not affiliated with any university, scientific establishment, or corporation, and therein lies his strength. Untainted by institutional bias or conventional wisdom, he brings fresh insight to the study of the ice ages. In the trailblazing tradition of James Croll, the self-educated Scotsman who discovered the link between ice ages and equinoctial precession; and Michael Faraday, the self-taught English physicist and pioneer in electromagnetism, Felix sounds an alarm that must be heeded. Ignore it at your peril.

Sugarhouse Publishing, P.O. Box 435, Bellevue, WA 98009-0435
To order call TOLL FREE 1-800-310-1764, ext 9761
In Washington call (206) 451-9311